Discovering Cosmetic Sc

Discovering Cosmetic Science

Edited by

Stephen Barton
Skin Thinking Ltd, UK
Email: stevebarton@skinthinking.com

Allan Eastham
Cosmarida Ltd, UK
Email: aeastham@cosmarida.co.uk

Amanda Isom
Bloom Regulatory Ltd
Email: amanda.isom@bloomregulatory.com

Denise McLaverty
Venture Logic Ltd, UK
Email: denisemclaverty@hotmail.com

and

Yi Ling Soong
The Body Shop International Ltd, UK
Email: yiling.soong@thebodyshop.com

Print ISBN: 978-1-78262-472-1
EPUB ISBN: 978-1-78801-713-8

A catalogue record for this book is available from the British Library

© The Royal Society of Chemistry 2021

All rights reserved

Apart from fair dealing for the purposes of research for non-commercial purposes or for private study, criticism or review, as permitted under the Copyright, Designs and Patents Act 1988 and the Copyright and Related Rights Regulations 2003, this publication may not be reproduced, stored or transmitted, in any form or by any means, without the prior permission in writing of The Royal Society of Chemistry or the copyright owner, or in the case of reproduction in accordance with the terms of licences issued by the Copyright Licensing Agency in the UK, or in accordance with the terms of the licences issued by the appropriate Reproduction Rights Organization outside the UK. Enquiries concerning reproduction outside the terms stated here should be sent to The Royal Society of Chemistry at the address printed on this page.

Whilst this material has been produced with all due care, The Royal Society of Chemistry cannot be held responsible or liable for its accuracy and completeness, nor for any consequences arising from any errors or the use of the information contained in this publication. The publication of advertisements does not constitute any endorsement by The Royal Society of Chemistry or Authors of any products advertised. The views and opinions advanced by contributors do not necessarily reflect those of The Royal Society of Chemistry which shall not be liable for any resulting loss or damage arising as a result of reliance upon this material.

The Royal Society of Chemistry is a charity, registered in England and Wales, Number 207890, and a company incorporated in England by Royal Charter (Registered No. RC000524), registered office: Burlington House, Piccadilly, London W1J 0BA, UK, Telephone: +44 (0) 20 7437 8656.

Visit our website at www.rsc.org/books

Printed in the United Kingdom by CPI Group (UK) Ltd, Croydon, CR0 4YY, UK

Foreword

After 23 years as a biochemist in academia, I joined Anita Roddick at The Body Shop, where I was shocked to realize just how much I needed to learn about the cosmetic industry. Despite my previous appointment, lecturing at Reading University in food science and technology, I discovered I was a real novice at cosmetic science. Being familiar with the constraints governing the food industry had not prepared me for all the nuances of cosmetics and personal care. For instance, I had no idea that cosmetics, unless stated otherwise, are formulated to have shelf-lives of up to three years! At that time, the focus in food research was on extending food shelf-lives to weeks and months. Yes, canned food can be stored safely for years, but cosmetics do not come in cans. Personal care formulas have a closer resemblance to food oils, flours and food emulsions, all of which had shelf-lives of, at best, up to a year.

While I wondered how this three-year stability was achieved, I slunk home and disposed of the no longer fashionable colour cosmetics that I had stashed away for far more years than I care to admit in case they became popular again. I am very grateful to my colleagues, the members of The Society of Cosmetic Scientists, and especially to the industry experts, many of whom are editors and authors of this excellent book, who taught me all I needed to pursue a rewarding career in cosmetic science.

Discovering Cosmetic Science
Edited by Stephen Barton, Allan Eastham, Amanda Isom,
Denise McLaverty and Yi Ling Soong
© The Royal Society of Chemistry 2021
Published by the Royal Society of Chemistry, www.rsc.org

I am proud that during my time at The Body Shop I played a significant a part in introducing hemp into mainstream cosmetics and by acquainting cosmetic formulators with food practices, The Body Shop team went on to develop the first body butters. Later, through the Society of Cosmetic Scientists, I met Steve Barton, who is the senior editor of this book, respected consultant to the industry and inspirational Lecturer in Cosmetic Science at The London College of Fashion. When I met Steve, he was Skincare Scientific Advisor at Boots and their No. 7 Protect and Perfect serum was hitting the headlines. Twenty-four hours after a TV programme reporting on the independent scientific studies on Boots No. 7 Protect and Perfect serum had been aired, national newspapers were saying 'not a jar was left in UK shops' and describing Steve as *'the hottest male property in Britain right now. Women have gone potty for his anti-ageing skin cream'*. I know Steve would much rather be known for his expertise in skin moisturization, his long experience working for Boots and Oriflame and for his many publications, including this book.

Discovering Cosmetic Science is not just another textbook but more an informative journey, which takes the reader through the most important and interesting aspects of cosmetic science. In the Introduction, Steve Barton with co-editors Allan Eastman, Amanda Isom, Denise McLaverty and Yi Ling Song modestly describe *Discovering Cosmetic Science* as a novice's guide, and certainly I would have found this book invaluable when I first joined the industry. Each author is expert in the science underpinning the different aspects of the cosmetics. Together they explain and illustrate this for the reader using separate boxes of text and images to allow the reader to explore deeper explanations or interesting scientific facts.

The 'route of this journey' has been chosen, deliberately, to build up the reader's appreciation of the science of everyday products – probably in the order they would use them.

Ed Rolls explains the fundamentals of cleansing products before Paul Cornwell describes the intricate structure of hair, and how this applies in hair care to ensure that bad hair days are a thing from the past. Steve Mason does the same for the equally complex and more delicate area of the mouth. Having showered, washed our hair and cleaned our teeth, Robin Parker leads the reader through the science behind skin and skin-care products. My own doctorate was on collagen synthesis so I understand skin physiology; however, good formulators must know *much* more. The reader discovers that, yes, skin gives protection, regulation and sensation, but that it is also a dynamic canvas, a continually moving *escalator of specialized skin cells* plus responsive surface microbiome, the importance of which we are only just beginning to appreciate. Skin-care products are also engineered to have a delightful feel to complement the improved looks, which, as the charity Look Good Feel Better knows well, can contribute hugely to people's lives. Claire Summers also applies her expertise to working on skin's dynamic canvas. Explaining the science of colour perception and creation of visual effects, Claire shows the reader that the science behind colour cosmetics is much more than a smudge.

Over the years, I have noticed how so many of us automatically smell products, on first use or as a test. If we perceive the odour as bad, then the product will not be used, irrespective of how delightful it feels or how well it performs. Virginie Danau, with her years of expertise in fragrances, has readers following the scent as she explains why fragrance is so important.

By now the average user, having completed some or all of this start-of-day routine, may get into some contentious discussions on social media about the latest 'must have' or 'avoid at all costs' ingredients. The final three chapters deal with facts to arm the reader for these types of discussion. First, Tony Causer outlines, in plain language and without marketing hype, the mechanisms behind many commonly discussed 'active ingredients'. By describing their chemistry and how to formulate to maintain their efficacy, he shows how careful formulators need to be in designing products that make a difference.

Stephen Kirk follows this up by explaining the testing that is essential to ensure that cosmetics are safe and efficacious before they are released onto the market. Testing to ensure stability – as I have already mentioned, many products can be expected to be stable for up to three years – is only part of the picture. Continual post-marketing surveillance of products ensures that this testing can be monitored against real-world use of products. Finally, last but certainly not least, Emma Meredith, Director-General at the CTPA, puts perspective on the many myths and scares to which our industry is subject, reassuring consumers that they can be confident in our highly regulated cosmetics.

Without realizing it, readers have been taken through the broad definitions of a cosmetic – cleansing, altering the appearance and odour – and the legal frameworks that apply to making cosmetics.

With this book, I am sure you will find the answers you were looking for and will enjoy your journey through the science behind an industry whose goal is to create delightful products that can clean, fragrance, oil, polish, protect and decorate. The industry's aim has always been to improve the health and well-being of everyone, and I like to think that cosmetic scientists are early adopters of the latest science and technology, adapting them for their own special requirements. Twenty plus years after joining Anita at The Body Shop, I find myself applying the very latest DNA technology, blockchain and AI to secure transparency in the cosmetic industry's complex supply chains. I regularly chair international cosmetic summits where experts from other sciences share their knowledge with cosmetic formulators, and I travel the world to learn and advise brands on cosmetic materials. If this industry has been a secret, hidden in plain sight from you, then welcome. Let this book be your way into cosmetic science.

Barbara Brockway
Past President of The Society of Cosmetic Scientists

Acknowledgements

The Editors are truly grateful for the support they have had from The Society of Cosmetic Scientists (SCS) and the Royal Society of Chemistry (RSC) in bringing this science to the attention of you, the public.

From its beginnings to its publication date, the SCS Council has provided great support and feedback under the guidance of the Presidents and other officers: Grace Abamba, Ruth Borner, Stewart Long, Mary Lord, Emma Meredith, Mojgan Moddaresi, Jackie Searle, Barry Winslett, Mustafa Varcin.

We would also like to thank Gem Bektas, SCS General Secretary, for her unstinting support and for handling the essential contractual and logistical aspects of putting this book together.

All of the Editors are members of the SCS Education Committee and we wish to acknowledge the continued support and encouragement from our colleagues over the last 2 years: Pauline Ayres, Daniel Burney Heather Carolan, Russell Cox, Anke Ginzburg, Susan Hurst, Lauren Kempen, Daksha Patel, Ian Prendergast, Lorna Radford, Claire Summers, Roger Rowson: Joyce Ryan, Prof. Danka Tamburic, Prof. Peter Taylor.

We acknowledge with particular thanks the support from other members of the Education Committee: Dr Barbara Brockway – for stimulating conversations that kicked off the whole idea and her continued support; Dr Tony Morton – for his

---

Discovering Cosmetic Science
Edited by Stephen Barton, Allan Eastham, Amanda Isom,
Denise McLaverty and Yi Ling Soong
© The Royal Society of Chemistry 2021
Published by the Royal Society of Chemistry, www.rsc.org

insightful comments on the penultimate drafts; Dr Bob Hefford – for providing information and critique for on-hair colourants; Bernice Ridley – SCS Education Programme Coordinator, for liaising between the authors and the editorial team, assisting in sourcing images, accessing and checking consistency with the Society's Diploma Course content, and proof reading.

At the RSC, Janet Freshwater and Katie Morrey proved to be invaluable in guiding us through this, our first experience of the publication process. We could not have achieved it without them. Whether assisting with sourcing and obtaining permissions to use images to illustrate the science in the book, or working with us on the front cover with their design team, through to ongoing support and guidance on the publicity, they have been a great part of our team.

Last, but not least, huge thanks are due to our authors, many of whom are writing for the lay public for the first time. Their responsiveness and flexibility and their lack of complaints as we suggested amendments and reminded them of the deadlines are a testament to their passion for sharing the science in the brilliant field of science we all share.

Thank you!
Steve Barton
Allan Eastham
Amanda Isom
Denise McLaverty
Yi Ling Soong

Contributor Biographies

Steve Barton
A skin biologist who, after more than 13 years' experience in academia, moved to the cosmetic industry 30 years ago, first with Boots UK and later Oriflame. Steve now consults on skin-care formulation design, claims testing and product communication. He has published widely and contributed to working groups on cosmetic claims with UK and European industry bodies. He is a Past President and Fellow of the SCS and formerly 'Practitioner in Residence' at the University of the Arts, London.

Tony Causer
Tony is Operations Director of Adina Cosmetic Ingredients Ltd – a technically focused distributor of personal care actives in the UK market. He began his career in the beauty industry when he was 17, starting in production, making creams and lotions, and then progressing to formulating as a chemist.

Paul Cornwell
Paul is a technical consultant and business development manager for TRI Princeton, a leading cosmetic science research institute based in New Jersey, USA. Paul is a qualified pharmacist and experienced cosmetic chemist with over 23 years' experience working in the cosmetics industry, where he has developed an

extensive knowledge of hair biology, hair fibre science and hair product science.

Virginie Daniau
After many years in the fragrance industry, working with Quest and Givaudan, Virginie is an Independent Fragrance Specialist. She has acquired extensive knowledge of this fascinating world, having developed successful fragrances for many top brands. Virginie retains a passion for perfumes and continues to share her wide experience with large corporations and private customers.

Pauline Dubois
Pauline qualified as a chemist specializing in formulation. For the past 10 years she has worked in her native France and the UK developing her skills in make-up and colour cosmetics. She is now a Senior Cosmetic Chemist working for C&R Packers in New Zealand.

Allan Eastham
Allan has had over 30 years' experience in the cosmetic industry, specializing in the past with soaps and detergents. In more recent times he has been a Cosmetic Safety Assessor with SGS and is now Technical Manager for Cosmarida with a more diverse specialization in self-tanning products and hair colourants. Diplomas in Quality Assurance and Packaging have helped him gain a good insight into the cosmetics industry.

Amanda Isom
Amanda is currently Regulatory Affairs Associate Director with Bloom Regulatory. Previously she worked as Compliance Manager for the Cosmetic Toiletry and Perfumery Association (CTPA) and has also held the position of Honorary Education Secretary for the Society of Cosmetic Scientists (2019–2020). Amanda has over 20 years' experience in the regulation of cosmetic products and related areas, knowledge that she shares with others as an author and examiner for the SCS Diploma in cosmetic science.

Stephen Kirk
Stephen is a highly qualified cosmetic regulatory professional with more than 20 years' experience working for Boots UK. His experience of New Product Development and prototyping has involved work with developers in Europe and the Far East.

This expertise, and his background in toxicology, have been invaluable in his work with CTPA, Cosmetics Europe and dermatologists on various regulatory issues. He is founder and owner of SK-CRS consultancy.

Jasmine Lim
Jasmine has an MSc in Cosmetic Science from the University of the Arts, London. After time in the fragrance and flavour industry, she joined the Good Housekeeping Institute. Here she carries out independent assessments of products, combining consumer feedback with laboratory testing, to find the best products on the market. Passionate about consumer education, she writes review articles for the *Good Housekeeping* magazine and website to help clarify the myths and misconceptions surrounding cosmetics.

Stephen Mason
Steve joined GSK Consumer Healthcare in 2005 as Medical Director for Sensodyne. Originally a PhD chemist, he has been a respected member of the oral health research community for almost 30 years, previously at Colgate Palmolive, Quintiles and Hill Top Research. He is a member of ORCA (European Organisation for Caries Research), IADR (International Association for Dental Research) and EADPH (European Association of Dental Public Health). Steve has published over 40 oral health research papers and conference presentations, many on the design, performance and benefits of oral healthcare products on consumers'/patients' quality of life.

Denise McLaverty
Denise has spent a number of years in the fragrance industry working with IFF and then as an independent consultant at Robertet. She has an MSc in Chemical Research and is passionate about the value of science education. Denise is an active member of the SCS Council and is currently Chair of the 'Scrub Up On Science' programme for 11–16-year-old students. This uses cosmetic science to illustrate some practical applications of the school science curriculum.

Emma Meredith
Emma is Director-General of the Cosmetic, Toiletry and Perfumery Association (CTPA). As Director-General she is

responsible for the strategic direction of CTPA and acts as the public voice of the Association. Emma is leading the Association's work focusing on the UK's exit from the EU, 'Brexit', and the future UK landscape, external stakeholder engagement and international relations. Emma is a pharmacist by profession.

Chris Metcalfe
Chris is Managing Director of Adina Cosmetic Ingredients Ltd – a UK distributor with a focus on active ingredients. He studied Business Studies at Sheffield Hallam University and went on to work in sales for a contract manufacturer of medical implants before joining Adina in 2011.

Robin Parker
Robin is the Technical Director of Acheson & Acheson, now part of The Hut Group, a UK manufacturer of premium skin-care products and toiletries. Robin began his career as a formulator with Max Factor and has subsequently worked with a number of UK-based contract manufacturers developing and supplying products to most of the UK's high-street retailers and many well-known brands. Robin is a Past President of the SCS and currently sits on two of the CTPA committees.

Rachael Polowyj
Rachael graduated from the University of the Arts, London, with an MSc in Cosmetic Science and presented her thesis as a poster at the International Federation of Cosmetic Chemists (IFSCC) Congress, Munich, 2018. She is a keen freelance beauty writer, focusing on trends and ingredients. Rachael currently works as an Account Manager for IMCD UK Ltd.

Eleanor Roberts
Eleanor's research at King's College London (KCL) and The Scripps Research Institute in La Jolla, CA, USA, was originally into neuroimmunology. In 2005, a love of writing led her into a career as a medical/science writer. Writing for numerous companies, academic institutions and charities in the healthcare sector, Eleanor explored a wide range of biosciences. Her interest in dentistry and cosmetic science arose through working with KCL's London Dental Institute. She developed an ongoing collaboration with GSK Consumer Healthcare and a fascination

with the chemistry of teeth and oral care. Eleanor is a member of the Association of British Science Writers.

Nichola Roberts
Nichola is a Senior Laboratory Technician at Acheson & Acheson, now part of The Hut Group. Nichola graduated with a master's degree in Cosmetic Science from the University of the Arts, London, in 2018. Since graduating, Nichola has been working within the innovation team at Acheson & Acheson, utilizing the latest trends and technologies to develop bespoke products for a variety of skin-care brands.

Edward Rolls
Ed has broad experience in many areas of the cosmetic industry, working for contract manufacturers, raw material manufacturers and distributors. With more than 15 years' experience in formulation, manufacturing and sales, his speciality areas are natural body care, prestige and budget skin care and surfactants.

Yi Ling Soong
Yi Ling has built up an extensive knowledge of product development and formulating, with over 6 years of experience. She joined The Body Shop as a Product Development Technologist in 2020 and was previously working as a Senior Development Chemist at Orean Personal Care Ltd, a contract manufacturer. She achieved a 1st class degree in Pharmaceutical and Cosmetic Science and completed the SCS Diploma in Cosmetic Science with Distinction, and was also awarded the SPC Prize for Best Essay in 2017.

Claire Summers
Claire is Technical Development Manager at Azelis Life Sciences, where she is responsible for the UK personal care laboratory. With over 25 years' experience in developing colour products, Claire shares this knowledge by contributing to the SCS Diploma course by being a member of the Education Committee of the SCS, as well as lecturing on the MSc in Cosmetic Science course at the University of the Arts, London.

Contents

Chapter 1
Introduction   1

Steve Barton, Allan Eastham, Amanda Isom, Denise McLaverty,
Yi Ling Soong and Rachael Polowyj

1.1	How Many Cosmetic Products Do You Use in the Day?	2
1.2	What Is a 'Cosmetic'?	3
1.3	What Goes into a Cosmetic Product?	5
1.4	The Importance of the Identities and Structures of Chemical Compounds	10
	1.4.1 Organic and Inorganic Chemistry	11
	1.4.2 Carbon Chains and Carbon Rings	11
1.5	The Importance of How Chemical Compounds Are Held Together	12
	1.5.1 Did You Know That Some Ingredients in Cosmetic Products Have an Electrical Charge?	12
	1.5.2 Did You Know That Some Ingredients in Cosmetic Products Have 'Polarity'?	12
	1.5.3 Did You Know That Some Compounds Are Held Together More Strongly Than Others?	13

Discovering Cosmetic Science
Edited by Stephen Barton, Allan Eastham, Amanda Isom,
Denise McLaverty and Yi Ling Soong
© The Royal Society of Chemistry 2021
Published by the Royal Society of Chemistry, www.rsc.org

1.6	The Importance of How Chemical Compounds Fall Apart	13
1.7	Prepared to Read On?	14
Reference		18

Chapter 2
Clean Chemistry – The Science Behind Cleansing Products 19

Edward Rolls

2.1	Introduction to Surfactant Behaviour		19
	2.1.1	So, What Makes Surfactants So Special?	19
	2.1.2	How Does the Special Structure Affect How Surfactants Behave?	20
	2.1.3	Why Isn't Using Water Alone Enough to Clean Things?	21
	2.1.4	Experiment at Home	23
	2.1.5	How Do Surfactants Create Foam?	23
	2.1.6	How Do Surfactants Help to Clean Dirt Away?	25
2.2	Surfactants Used for Cleaning		26
2.3	Secondary Surfactants – Luxurious, Creamy Foams		31
2.4	Finishing Touches		35
2.5	Alternative Systems		37
2.6	Natural/Organic and Sustainability		38
2.7	Conclusion		42

Chapter 3
Good Hair Day: The Science Behind Hair-care Products 43

P. Cornwell and J. Lim

3.1	Introduction	43
3.2	Hair Structure	44
3.3	Hair Diversity	52
3.4	The Living Follicle	54
3.5	Sebum and Hair Greasiness	57
3.6	Hair Damage	58
3.7	Hair Thinning and Hair Loss	60
3.8	Hair Greying	61
3.9	Shampoo Surfactant Bases	62
3.10	Shampoo Conditioning Systems	64

Contents   xix

3.11 Hair Conditioners   66
3.12 Hair Styling   69
3.13 Straightening Treatments   70
3.14 Bringing It All Together   73
References   73

Chapter 4
Oral Care – A Mouthful of Chemistry   75

Eleanor Roberts and Stephen Mason

4.1 Physiology of Teeth   75
    4.1.1 Overview and Structure   75
    4.1.2 Enamel and Dentine   77
    4.1.3 Saliva   77
    4.1.4 The Pellicle   78
4.2 When Good Mouths Go Bad   79
    4.2.1 Plaque (aka Dental Biofilm)   79
    4.2.2 Dental Calculus (aka Tartar)   80
    4.2.3 Periodontal Disease (Gum Disease)   80
    4.2.4 Tooth Decay (aka Dental Caries)   82
    4.2.5 Tooth Wear (aka Dental Erosion)   84
    4.2.6 Dentinal Hypersensitivity (aka 'Sensitive Teeth') 85
    4.2.7 Tooth Stain and Whitening   87
    4.2.8 Oral Malodour (Bad Breath)   88
4.3 Just What Are All These Ingredients in My Oral Care Product?   88
    4.3.1 Overview   88
    4.3.2 Fluoride   90
    4.3.3 Abrasives   93
    4.3.4 Anti-sensitivity   94
    4.3.5 Stain Removal/Whitening   97
    4.3.6 Anti-gingivitis   100
    4.3.7 Calculus Control   102
    4.3.8 Enamel Care   102
    4.3.9 Other Ingredients   102
    4.3.10 Mouthwash   104
    4.3.11 How Are Claims for Oral Care Product
           Performance Substantiated?   104
4.4 Conclusion   107
Further Reading   107
References   108

Chapter 5
You Against the World! – The Science Behind Skin and Skincare Products   109

Robin Parker, Nichola Roberts and Monique Burke

5.1	The Skin – What Exactly Does Our Skin Do?	109
	5.1.1 Skin Deep – What Is Beneath the Surface?	110
	5.1.2 We Are Not Alone – What's On The Surface?	114
5.2	One Size Fits All? – All Skin Is Different	116
	5.2.1 How and Why Does the Skin's Appearance Change with Age?	116
	5.2.2 Does Skin Vary from Individual to Individual?	120
	5.2.3 Are There Differences Between Men's and Women's Skin?	122
5.3	Staying on the Surface – Do Cosmetic Ingredients Go into The Skin?	124
5.4	What Goes into Skincare Products and Why – Care and Protection for Your Skin	126
	5.4.1 How Do Cosmetics Make a Difference to Skin Appearance?	127
5.5	Moisturization – Where Chemistry Meets Biology	129
	5.5.1 What Is the Difference Between Moisturization and Hydration?	129
	5.5.2 Which Ingredients Are Important in an Effective Skin Moisturizer?	131
5.6	Emulsions – Better Together!	132
	5.6.1 What Are Emulsifiers and Why Are They All Different?	135
	5.6.2 How Can We Make Emulsions Stable for Several Years?	138
5.7	Touch and Texture – It's Just A Feeling	141
	5.7.1 What Is Rheology and Why Is It So Important in Skincare?	142
	5.7.2 How Can We Control the Rheology and Skin Feel of Cosmetic Creams?	144
	5.7.3 By What Other Ways Can We Affect the Feel of Skincare Formulas?	146
5.8	Different Types of Skincare Products	148
	5.8.1 Why Do We Have Day and Night Moisturizers, and Are They Different?	148
	5.8.2 Do Men Need Different Moisturizers to Women?	149

	5.8.3 How Important Are Skincare Regimes and What Are the Necessary Products?	149
	5.8.4 Are the Skin Concerns for the Body the Same As Those for the Face? – How Many Different Moisturizers Do We Need?	152
5.9	Conclusion	153
References		153

Chapter 6
More Than a Smudge of Colour – The Science Behind Colour Cosmetics — **155**

Claire Summers and Pauline Dubois

6.1	Why Does Something Appear Coloured?	155
6.2	How Can We Create Coloured Products?	161
	6.2.1 Dyes	162
	6.2.2 Why Pigments Are Crucial for Colour Cosmetics	163
	6.2.3 Are Inorganic and Organic Pigments the Only Materials Used to Create Colour?	168
6.3	Are Pigments Easy to Use?	173
	6.3.1 How Do You Disperse Pigments?	174
6.4	Why the Texture of Cosmetic Formulations Is so Important	177
	6.4.1 Foundations – Are They More Than Just Colour?	178
	6.4.2 Powders – Simple Yet Surprisingly Complex	181
	6.4.3 Lipsticks – More Than Lip Service	183
6.5	The Art of Colour Matching	184
6.6	Curl up and Dye?	184
	6.6.1 Oxidation and pH	187
	6.6.2 Natural Dyes	190
	6.6.3 Temporary Hair Colourants	190
6.7	Conclusion	191
Further Reading		191

Chapter 7
Follow the Scent – The Science Behind the Fragrance in Products — **192**

V. Daniau

7.1	Sources and Mechanism of Odour Formation	192
	7.1.1 Mechanisms	193

7.2	Odour Recognition (Box 7.1)		194
	7.2.1	Specialized Olfactory Sensory Neurons	194
	7.2.2	Infinite Combinations	194
7.3	Smell and Emotions		195
7.4	The Message Carried by a Fragrance		196
7.5	Creating and Masking Odours		197
	7.5.1	The Creative Process and the Teams Involved	197
	7.5.2	The Construction of a Fragrance	201
	7.5.3	Fragrance Families	204
	7.5.4	Fragrances in Different Bases and Products	206
	7.5.5	Covering Malodours	208
7.6	Stability: Why Do Fragrances Change Over Time?		208
	7.6.1	Base Interaction	208
	7.6.2	Stability	209
7.7	Essential Chemistry		212
	7.7.1	How Were Aroma Chemicals Discovered?	212
	7.7.2	Categories of Aroma Molecules	215
7.8	Extraction Methods		216
	7.8.1	Expression	216
	7.8.2	Distillation	219
	7.8.3	Solvent Extraction	221
	7.8.4	Developments in Scientific Research Methods to Analyse Natural Scents	222
7.9	Conclusion		223
Further Reading			224

Chapter 8
The Inside Story – The Science Behind Active Ingredients   225

C. Metcalfe and T. Causer

8.1	Vitamins		226
	8.1.1	Where Do Vitamins Come From?	226
	8.1.2	Oil-soluble Vitamins	227
	8.1.3	Water-soluble Vitamins	229
	8.1.4	Minerals	231
	8.1.5	Other Vitamins	231
8.2	Peptides		231
	8.2.1	Why Are Peptides Useful in Cosmetics?	232
	8.2.2	How Are Peptides Named?	233

	8.2.3	Discovering New Peptide Ingredients	233
	8.2.4	Which Peptides Are Commonly Used?	234
	8.2.5	Is It Just Hype?	234
8.3	Hydroxy Acids	235
	8.3.1	Why Are Hydroxy Acids Useful in Cosmetics?	236
	8.3.2	Hydroxy Acids in Peel Products	236
	8.3.3	Using Products Containing AHAs	237
8.4	UV Filters: Protecting Products and the Skin/Hair	237
	8.4.1	The Electromagnetic Spectrum	237
	8.4.2	What Is Sun Protection Factor (SPF)?	238
	8.4.3	How Does UVA Protection Differ from SPF?	239
	8.4.4	How Do Sunscreen Products Work?	239
	8.4.5	Product Innovation	240
	8.4.6	Why Do Coloured Cosmetics Sometimes Contain UV Filters Even If They Don't Offer UV Protection?	240
	8.4.7	The Hair Needs Protecting Too!	240
8.5	Antioxidants	241
	8.5.1	What Is Oxidation?	241
	8.5.2	How Do Antioxidants Work?	244
8.6	Antimicrobials	245
	8.6.1	The Germs (Microorganisms) Around Us	246
	8.6.2	Products Need Protecting – 'Preserving'	250
	8.6.3	What Happens if Cosmetics Are Not Preserved?	251
	8.6.4	How Do Companies Know If Their Products Will Remain Safe If They Become Contaminated?	251
	8.6.5	Antimicrobial Protection on the Skin	253
8.7	Natural Extracts	253
	8.7.1	Producing Natural Extracts Using Stem Cells	256
8.8	Delivery Systems	256
	8.8.1	Why and Where Are Delivery Systems Used?	257
	8.8.2	Examples of Delivery Systems	258
8.9	Antiperspirant and Deodorant Effects	259
	8.9.1	What Is the Difference Between an Antiperspirant and a Deodorant?	259
	8.9.2	Why Do Antiperspirants and Deodorants Come in Different Formats?	260
8.10	Conclusion	261

Chapter 9
Testing and More Testing – The Science Behind Keeping Your Skin Safe and Healthy   262

Stephen Kirk

9.1	Cosmetic Products – How We Keep You and Your Skin Safe and Healthy	262
9.2	Stability Testing – Making Sure a Product Is Fit for Purpose	263
	9.2.1 Microbiological Testing – Will It Go Mouldy?	264
	9.2.2 Chemical Stability – Will My Product Change Colour?	269
	9.2.3 Period After Opening (PAO) and Shelf-life – How Long Will It Be Okay to Use?	272
9.3	Safety Assessment of Cosmetic Ingredients and Finished Products – Is My Chosen Product Going to Harm Me?	273
	9.3.1 Hazard and Risk – The Tale of a Shark and the Swimmer	273
	9.3.2 Identifying the Hazard Characteristics of a Cosmetic Ingredient	276
	9.3.3 Testing New Cosmetic Ingredients Using Non-animal Alternative Methods	277
9.4	Product Claims – Will My Product Do What It Says on the Tin?	278
	9.4.1 Sun Protection Testing – Will My Cream Protect Me from the Harmful Effects of the Sun?	278
9.5	Safety-related Claims Made on Some Products	284
	9.5.1 Will Your Product Damage My Eyes or Make Them Sting?	284
	9.5.2 Hypoallergenic – Exactly What Does This Mean?	285
9.6	Post-market Surveillance – the Customer's Story of Using a Product	287
9.7	What Does All This Mean to Me?	288
Further Reading		288

Chapter 10
Myths and Scares – Science in Perspective   290

E. Meredith and R. Polowyj

10.1	Are Cosmetics Tested on Animals?	291
	10.1.1 European Union (EU)	291
	10.1.2 Global Challenges	292

10.2	How Much Does the Skin Absorb?		292
	10.2.1	Myth – 60% of Everything You Put on Your Skin Is Absorbed	293
	10.2.2	Fact – Different Skin Types Have Different Barrier Functionality	293
10.3	Should I Avoid Certain Ingredients?		294
	10.3.1	Why Are Some Products Labelled As Being 'Free-from' Certain Ingredients?	294
	10.3.2	Parabens	296
	10.3.3	Sulfates	297
	10.3.4	Silicones	297
	10.3.5	How Do I Know That the Ingredients in the Products I Use Are Safe?	299
	10.3.6	Do Cosmetics Contain Hormone-disrupting Ingredients?	300
10.4	What Is the Difference Between Natural and Synthetic Ingredients?		301
10.5	Do Cosmetics Pollute the Oceans?		304
	10.5.1	Do Cosmetics Contain Plastic Microbeads?	306
	10.5.2	Are Cosmetics a Cause of Microplastics?	306
	10.5.3	Why Is Plastic Packaging So Often Used to Package Cosmetics?	307
	10.5.4	Are Sunscreens Damaging to Coral?	309
10.6	Why Are Cosmetics So Important?		309
10.7	Having Confidence in Cosmetic Products		311
References			312

Appendix 1 — 313

Subject Index — 317

Reviews — 339

CHAPTER 1

Introduction

STEVE BARTON,*[a] ALLAN EASTHAM,[b] AMANDA ISOM,[c] DENISE MCLAVERTY,[d] YI LING SOONG[e] AND RACHAEL POLOWYJ[f]

[a] Skin Thinking Ltd, Nottingham, UK; [b] Cosmarida, Sheffield, UK; [c] Bloom Regulatory Ltd, London, UK; [d] Venture Logic Ltd, UK; [e] The Body Shop International Ltd, UK; [f] IMCD UK Ltd, UK
*Email: stevebarton@skinthinking.com

Welcome to this 'novice's guide' – at last, a book that explains the real science behind the cosmetics you use. We are assuming nothing about you, the reader, your background or expertise. The fact that you are reading this introduction suggests that you are interested in finding out more! Which is why we had the idea of putting these chapters together. We guessed that there may be quite a few of you wanting to find out what some of those strange-sounding chemicals listed on your shower gel are doing in a product, and why. Or maybe you want to settle an argument about what cosmetics actually do. We'd like to think that there are lots of teachers or journalists needing to know more about a subject you'd like to communicate to others. Students or others

---

Discovering Cosmetic Science
Edited by Stephen Barton, Allan Eastham, Amanda Isom,
Denise McLaverty and Yi Ling Soong
© The Royal Society of Chemistry 2021
Published by the Royal Society of Chemistry, www.rsc.org

wanting to find out if this is the career for you – we'd love to encourage you by sharing the authors' many years of scientific knowledge. Or you may simply like the front cover! Whatever the case, we have plenty of science for you to discover. This introductory chapter aims to set the scene and start you thinking about the science and concepts that you'll come across in the subsequent chapters of this book. Let's start by asking you a question.

1.1 HOW MANY COSMETIC PRODUCTS DO YOU USE IN THE DAY?[†]

Do you shower first thing? Or clean your teeth? Or rub oil into your beard? Or put dry shampoo on your hair because you are in a hurry? All of these activities will involve some kind of cosmetic product. How do all these products work? Why might they be needed? Are they all 'hype'? To answer these and other questions, we'll try to use a gentle approach, giving you a 'guided journey' through the different product types.

You'll see from the Contents list that we have approached this journey by stopping off at each of the 'core technologies' and the product areas they support. Then we've added some important destinations – showing you the science behind safety – before ending up by dispelling some of the myths that find their way into the public domain. In reality, a cosmetic scientist developing a product will need a broad understanding of all of these topics in their daily work.

Reading this book, you'll discover that, while cosmetic products often sit on the surface, they are not as superficial as often thought. We think that you will learn that there's some amazing science behind them – we will go into this in some detail in places, but you won't need a PhD to understand the science. We shall also point out some interesting facts on our way, uncovering some of the truths behind the myths. Look out for Boxes in the chapters where authors either explain things in more detail or go off on an interesting side road during your journey of discovery. We've also pointed you in the direction of

---

[†]Most people say it's around 5 or 6 – soap, shampoo, toothpaste, deodorant and moisturizer. But Figure 1.1 shows that it could easily become 20 or more!

further reading if you want to know more. We know you'll have lots of questions to ask and we will try to pre-empt these if we can.

To set the scene, we suspect your next question may be ...

1.2 WHAT IS A 'COSMETIC'?

The word 'cosmetic' shares its origin with 'cosmos', from the Greek '*kosmos*', their word describing 'order' or 'ornament'. This second meaning has led to 'cosmetic' being used to describe the outer appearance. This focus on surface or superficial matters has also led to the interpretation that cosmetic *products* are somehow trivial or lacking depth; we think you'll change your mind after reading this book. This 'ornamentation' definition also leads many to think that the term cosmetic *product* just applies to 'makeup'. Makeup, sometimes called colour cosmetics or decorative cosmetics, is indeed one class of cosmetic product. However, there are many different classes used by millions of consumers every day. Why? Well, mostly to help keep themselves clean and hygienic, looking and feeling fresh and maintaining a healthy appearance. Generally, then, the definitions of cosmetics and personal care products around the world are similar. They focus on the 'appearance' of the outer body surfaces – the skin, hair and teeth – keeping them clean, perfuming or correcting body odours. Definitions also include *changing* the appearance of these surfaces, protecting them and keeping them in good condition, or, as in the other meaning of '*kosmos*', in good 'order'. It is these aims that all cosmetic science is designed to achieve.

Figure 1.1 gives some examples of the myriad of products covered by the term 'cosmetic'.

Each area of the world will have its own legal definition of 'cosmetic product'. Although there are many commonalities, some products classified as cosmetic in the UK and European Union (EU) are classed as different product types in other jurisdictions. For example, sunscreens, which are cosmetic products in the UK and EU, are classed as over-the-counter (OTC) medicines in the USA; hair dyes are cosmetic products in Europe and the USA but are classed as 'quasi-drugs' in Japan.

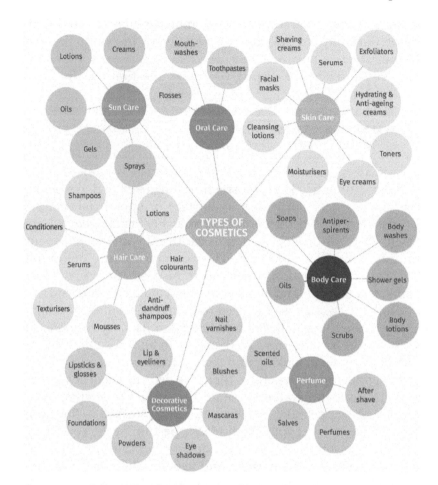

Figure 1.1 Diagram showing the range of types of cosmetic products organised and colour coded in their 'families'. This useful reference should help guide you through the coming chapters.
Reproduced with permission from Cosmetics Europe (https://cosmeticseurope.eu/cosmetic-products).

The main purpose of the laws governing the manufacture and sale of cosmetic products is consumer safety. Companies making and selling cosmetic products, and the regulatory bodies charged with implementing and policing the legislation, have a duty of care to ensure that human health is not compromised. You will find out some of the ways in which this works in practice, and the science behind it, in Chapter 10.

Introduction

However, wherever a cosmetic product is placed on the market in the world, the core science and technologies are the same.

1.3 WHAT GOES INTO A COSMETIC PRODUCT?

Hopefully, by now you are eager to know more about this; once you get into the later chapters, you'll find out more. However, there may be one feature of your cosmetic products you're already aware of – the list of ingredients, often seen on the back of a pack. This legal requirement is designed to help people identify, and so avoid, products with ingredients that they know they are sensitive or allergic to. Although there are minor differences around the world, the European markets implemented the International Nomenclature of Cosmetic Ingredients naming system, known as INCI, and this has also been adopted by many countries worldwide. INCI names are based on chemical identity and structure and you'll discover some of these as you read on (for more on naming conventions see Section 1.4). However, we will try not to overload you with too many of these. We can't possibly cover them all, but we've tried to give you some insights into what is used and why (see Box 1.1). So here we can start by talking about the different roles that ingredients play in the product.

When cosmetic scientists talk to each other about ingredients, they may use terms such as 'base', 'workhorse' or 'chassis' to describe some ingredients. We suspect you'll guess what these are – the unsung heroes of every product. In Chapter 2 you'll be introduced to one such class – surfactants – and learn what these are and why this class of ingredients is so important; so important that they appear again in Chapters 3 and 4 and, although they get called emulsifiers, again in Chapters 5 and 6 – yes, we know we said this would be gentle; stick with us!

There are many other classes of ingredients that are used across all types of product. Some, such as polymers, may be there to stabilize a product but also contribute to the 'feel' of the product in use. Others, such as glycols and alcohols, may be there to keep an ingredient soluble in the product or give a particular feel on the skin. Yet others, such as fats and oils (common examples of chemicals known as 'lipids') used in cosmetics, can add lubricity – to help the product spread while in use – in addition to performing functions you'll discover in

BOX 1.1   INCI LISTINGS

Table 1.1 shows two simple products and their INCI listings. In this case we also show the function of the ingredients in the products. For several reasons it is not possible to show all this information on a product INCI listing – that's not the purpose of the listing, some ingredients may have more than one function and space on the label is often limited. Even these very simple products have a long list of ingredients that must be declared and, as you will see, many different functions.

Table 1.1   INCI listings for body wash and body lotion.

Body wash		Body lotion	
INCI name	Function	INCI name	Function
Aqua	Solvent	Aqua	Solvent
Sodium laureth sulfate	Surfactant	*Butyrospermum parkii* butter	Skin conditioning
Cocamidopropyl betaine	Surfactant	Dimethicone	Emollient
Sodium chloride	Viscosity controlling	Cetyl alcohol	Emollient/emulsion stabilizer
Phenoxyethanol	Preservative	Stearic acid	Emulsifier
Benzophenone-4	UV filter	C12–C15 alkyl benzoate	Emollient
Salix nigra bark extract	Skin protecting	Sorbitol	Humectant
Benzoic acid	Preservative	Phenoxyethanol	Preservative
Dehydroacetic acid	Preservative	PEG-100 stearate	Emulsifier
Citric acid	pH adjusting	Olus oil	Emollient
Hexylene glycol	Solvent	Glyceryl stearate	Emollient/emulsion stabilizer
Sodium citrate	Buffering, chelating	Caprylyl glycol	Preservative
Acid Violet 43	Cosmetic colourant	Magnesium aluminium silicate	Viscosity controlling
CI 42090	Cosmetic colourant	Carbomer	Emulsion stabilizer
CI 16035	Cosmetic colourant	Tocopheryl acetate	Antioxidant
CI 47005	Cosmetic colourant	Sodium hydroxide	pH adjusting
		Panthenol	Skin conditioning
		Olea europaea fruit oil	Emollient
		Alcohol	Solvent
		CI 42090	Cosmetic colourant

Many companies, however, *do* provide more information on their websites, describing some of the ingredients that they use and why.

We will not go into every item here – keep on reading this book and you'll find out more about the functions and chemistry of surfactants, emulsifiers, emollients, colourants, preservatives and so much more. However, there are a few important things to pick out, as follows.

Aqua – Water! Why is it given a Latin name? – well, it is short and commonly understood around European countries where INCI names were intended to be used. You will also notice here, and in the majority of products, that aqua is the main ingredient as a solvent.

More Latin names – *Salix nigra* bark extract; *Butyrospermum parkii* butter; *Olea europaea* fruit oil. These are all plant materials commonly used for many years in cosmetic products – willow bark, shea butter and olive oil, respectively. However, using the botanical (Latin) name for the plant *and* the part of the plant is important. Common names of plants differ around the world; different parts of a plant contain different chemical compounds, some of which could be toxic. This scientific naming convention ensures a universal understanding of the part of the plant being used and its safety profile.

Alcohol – You'll see this term twice in the body lotion. 'Alcohol' is class of chemicals that have a carbon backbone and an OH (hydroxyl) group. 'Alcohol' in an INCI list means ethyl alcohol (ethanol), with its solvent and cooling properties – chemically the same as the alcohol in wine, beer, *etc.* Cetyl alcohol is a 'bigger' alcohol in molecular size – it has more carbons in its chain – and that makes it a waxy solid rather than a liquid. Just to make life really interesting there are other names in these INCI lists that end in '-ol', *e.g.* phenoxyethanol; hexylene glycol; caprylyl glycol; panthenol. The '-ol' indicates that these chemicals have a hydroxyl group in their structure. For those of you who know Sweden, you may recognise the term 'öl', being Swedish for beer!

More than one function – You will see here that some ingredients are given more than one function in the product. One of the many wonders of chemistry is that one molecular structure can confer several properties – you only have to think of the many uses of water to understand this. Formulators choosing cetyl alcohol know that its waxy nature leaves a softening, emollient effect on the skin *and* that it also forms part of the emulsion structure. This can help stabilize the product – see Chapter 5 for more on this.

Another thing that you'll notice is that, unlike a cooking recipe, there is no indication of the proportions used. This is for good reason. First, it is not a recipe; INCI lists inform the user of anything they might want to avoid using. Second, companies want to keep their special creations to themselves – in some cases this goes as far as patenting their intellectual property. However, companies follow two conventions of INCI listing:

1. The ingredients used at greater than 1% are listed in order of concentration; those at less than 1% can be included in any order. A top tip when trying to assess which are the lesser ingredients is to look for common preservatives – benzoic acid; parabens; phenoxyethanol – or buffering chemicals – citric acid; sodium hydroxide – or fragrance (parfum). These are usually (but not always) used at less than 1%.
2. If you or anyone else reacts so badly to a product that they end up seeing a dermatologist, companies will release the full details, including levels of use, to help healthcare professionals assess what has happened. Once any intolerance or allergy has been diagnosed, the INCI list allows the individual to avoid this ingredient in the future.

Chapter 5. What this demonstrates is that one ingredient type can have a number of uses, even in the same product. So, unlike medicines, where an active ingredient, or ingredient combination, is responsible for the therapeutic effect, several cosmetic ingredients are blended together to provide a combination of sensorial and functional benefits. This is worth remembering whenever you hear or read a self-appointed 'expert' talking about the importance of 'the latest must-have *ingredient*' – if it isn't in an appropriate formula it may not be doing what you are told it is doing. This is why testing the finished product is important – as you'll discover in Chapter 9.

On the topic of 'active ingredients', Chapter 8 will explain the role of some common examples. Here again, the term 'active' is how cosmetic scientists describe them, to differentiate them from the 'chassis' ingredients. Once the press hear cosmetic scientists talk with pride about their latest product, with the latest 'active ingredient', this term gets into public conversation. We suspect that if you've stuck with us this far you've probably heard the term 'active ingredient' yourself! Once again, remember that it is the whole product you are using, not simply one ingredient, and a lot of thought and scientific understanding has gone into the combination of ingredients in the products you use.

We think that you will also quickly discover while reading these pages that it is not just about the science of *ingredients* and the *products*. The science behind skin, hair, nails and teeth also goes into the product. Understanding how the biology and chemistry interact is an important part of the process of creating a cosmetic product. When you then add in the physics – for example, how light illuminates and bounces off the skin and hair to create their familiar colour and appearance – you'll begin to understand how many aspects of science are covered in this book. You may even rediscover facts about science that you learned in school or college.

As a final thought about 'what goes into a product', it may not have escaped your notice that words such as 'looking' and 'feeling' keep cropping up. Despite some of the more extreme headlines announcing (once again) the best 'miracle anti-ageing moisturizer', all cosmetic scientists recognize that users look for the sensation of using a product as much as they look for 'an effect'. In fact, these 'sensorial' properties – touch, taste, appearance, smell (and even sound!) – are important 'effects' in their own right when it comes to us choosing a product. A lot of work goes into this, from the selection of ingredients based on their physical chemistry and aesthetics to the testing of the product's sensorial effects in trials with consumers. The importance of sensorial properties cannot be understated – if a product wasn't pleasurable to use it might not stand a chance of achieving *any* effect – the user will quickly bin a product that feels awful. This is worth remembering as you read through this book; it will be mentioned from time to time but especially in Chapter 5, where you will discover why touch is so important in skin and skincare products. In Chapter 7, you'll discover the important role that fragrance can play in ensuring the acceptability of a product and in providing sensorial benefits in the product. We will introduce you to the basic science behind all these senses in this book, but the topic deserves a whole book to itself. For anyone wanting to find out more details on these properties, we suggest dipping into *Sensory Evaluation: A Practical Handbook.*[1]

1.4 THE IMPORTANCE OF THE IDENTITIES AND STRUCTURES OF CHEMICAL COMPOUNDS

Now for 'the science bits'. We promised that you wouldn't need a PhD to understand the scientific facts and concepts in this book; neither do we want to talk down to you. So, as part of our 'gentle approach', we'd like to confirm some of the common scientific terms that you'll come across, to make sure we are 'all on the same page'. We'll revisit many of these in certain chapters where necessary, sometimes in some detail, but here are some concepts and descriptions to start you off.

Scientists need a common language to ensure that *we* are all on the same page too. INCI names are part of that language but, at another level, chemical names communicate the identities and structures of compounds. There are no rules on how the INCI names are printed on pack (font, size, capitalization) or referred to in ingredient literature. Since this is essentially a science book, whether referring to chemicals identities or INCI names we have followed the convention for naming chemicals where capital letters are not commonly used. For the cosmetic chemist, the names also hint at the behaviours of the ingredients that we want to put into products. Ingredient solubilities, whether they are liquid or solid, how stable they are, what they may react with (intentionally or unintentionally) and sometimes their potential toxicity can often be deduced from their chemical names.

Introduction

1.4.1 Organic and Inorganic Chemistry

We can consider this to be the most fundamental distinction in chemistry. You may have heard about the Periodic Table of the Elements – in 2019 we celebrated 150 years since its inception. It shows how the atomic structures of the different elements help define their chemical behaviour.

Inorganic chemistry refers to these elements and their various permutations in chemical compounds such as common salt (NaCl) found on potato chips or silica (SiO_2) found in microprocessor chips.

Organic chemistry refers to complex combinations of carbon and hydrogen and their compounds also containing nitrogen, oxygen and chlorine and, less frequently, sulfur or phosphorus.

The majority of cosmetic ingredients are organic but, as you will see in Box 1.1 and on many occasions in this book, understanding the interactions between organic and inorganic materials is an essential skill for the cosmetic chemist.

The term 'organic' often creates confusion in communicating the science behind cosmetics. The 'organic movement' is a totally separate concept created by organizations wanting to describe crops farmed without the use of synthetic pesticides and fertilizers.

1.4.2 Carbon Chains and Carbon Rings

The structure of organic chemicals can be based on a backbone of chains of carbon atoms – such as those found in oils and waxes – or based on a carbon ring structure – such as sugars and cellulose. The molecular size – the number of 'carbon units' – and the complexity of their arrangements – for example, chain branching or 'substitution' of carbons by other atoms such as nitrogen and oxygen in the structure – give endless permutations of form and function. Fortunately, the naming conventions help us to understand how to use them effectively.

Why Is This Important? These structural factors have an impact on the physical properties and interactions with skin, hair and oral cavity surfaces, which are themselves complex chemical structures. On one level, they can determine an ingredient's sensory properties; on another level, structural factors can also determine an ingredient's interactions with the biological surfaces in a beneficial or detrimental way.

The forces holding the compounds together also play a role, which brings us to the importance of how chemical compounds are held together.

1.5 THE IMPORTANCE OF HOW CHEMICAL COMPOUNDS ARE HELD TOGETHER

All chemical structures, including those in skin, hair and oral cavity, are held together by a number of different forces. This 'bonding' between the different elements within a compound helps to hold the structures together. You may come across a number of these in this book. Here are just a few.

1.5.1 Did You Know That Some Ingredients in Cosmetic Products Have an Electrical Charge?

Let us take common table salt as an example. Salt (sodium chloride, NaCl) is made up of sodium and chlorine. Because the sodium part has a positive charge and the chloride part has a negative charge, they are attracted together (just like magnets). Because they have an electrical charge, they are known as 'ions'. The positive sodium ion is known as a 'cation' and the negative chloride ion as an 'anion'. When positive and negative charges attract they form strong bonds – just think how hard it can be to pull magnets apart! In cosmetic products, strong bonds can also be formed between ingredients and this can help to keep the product stable, or help it to do what it was designed to do! You'll read more about this in Chapter 2 when we look at the surfactants in cleansing products. If you look at INCI lists on product packs, you may often find common salt in your products.

Why Is This Important? Many body surfaces carry a charge – you only have to rub a balloon against your clothing, then hold it to your head and see how the hair stands up to understand this. Cationic materials in particular will be attracted to body surfaces, which is great if you want to leave a layer on the surface.

1.5.2 Did You Know That Some Ingredients in Cosmetic Products Have 'Polarity'?

Just like inorganic compounds can have ionic charge, many organic compounds can have 'polarity'. Some have no polarity at all – 'non-polar' – whereas others have strong polarity – 'polar'.

Introduction

Once again, these are like the poles of a magnet, with attraction between opposite poles and repulsion between similar poles. Although these forces are far less powerful than magnets, at small molecular distances they play important roles in many chemical and biological systems.

Why Is This Important? Polarity is important in many biological systems – for example, it determines how cell membranes function. A more practical implication for cosmetic science is that 'like dissolves like' – a polar liquid will dissolve a polar substance. Water and alcohol are common polar solvents in cosmetic products. Mineral oil is non-polar and can solubilize other non-polar substances. Many natural oils comprise mixtures of polar and non-polar compounds, offering the best of both worlds when it comes to solubilizing other oils. As you will learn later, many non-polar materials can be chemically modified to make them more polar – a process common in creating surfactants and emulsifiers.

1.5.3 Did You Know That Some Compounds Are Held Together More Strongly Than Others?

With so many different ways in which a compound can be held together, it is not surprising that some are more stable than others and some more reactive.

Why Is This Important? You will discover in Chapter 9 that products need to be stable and undergo testing to ensure that this is achieved. Before that testing happens, the cosmetic chemist needs to understand the relative stabilities or reactivities of the ingredients they are dealing with to reduce the risk of the products falling apart. A lot of the time, effort and expense can be saved by thinking ahead. Once a product has passed its stability test, a cosmetic chemist gets a great sense of achievement, and if there are failures, understanding why and how they can be prevented next time can be a very satisfying learning experience too.

Which brings us to the importance of how chemical compounds fall apart.

1.6 THE IMPORTANCE OF HOW CHEMICAL COMPOUNDS FALL APART

You will know from your own experience of food and drink that most things eventually 'go off'. Whether it is the browning of a cut apple or butter going rancid, Nature has a way of gradually falling apart. Both of these processes are examples of 'oxidation', a very important force in cosmetic science – whether it is helping to create desirable effects such as hair colouring or an important mechanism in the undesirable effects of skin ageing.

You will also know that there are ways of slowing these processes. For example, lemon juice slows the browning of a cut apple by an 'antioxidant' mechanism; you'll discover more in Chapter 8. Keeping butter in a refrigerator slows rancidity. Temperature is important in most chemical reactions, including the process of oxidation.

Why Is This Important? The choice of starting material is clearly an obvious way of preventing a problem in the first place, as was hinted at above. It is important to understand the factors that contribute to compounds falling apart – extremes of temperature, extremes of pH and ultraviolet light exposure are just a few of the factors. The basic chemical structure can be a guide – for example, unsaturated fatty acids, beneficial in the diet and in cosmetics, are more prone to oxidation than saturated fatty acids. It is important that cosmetic scientists understand how to control these potentially destabilizing factors to keep the product formulation in its intended state of quality during its lifetime of use. An essential part of this is understanding what happens to the ingredients during the manufacturing process. Simple things such as the order or timings of addition of an ingredient can make a real difference.

1.7 PREPARED TO READ ON?

In this introduction to cosmetic science, we have tried to prepare you for the scientific terms that you will come across in the following chapters, and many more will be explained as they arise in the text. If these explanations still leave you needing to know more, we hope that this means we have triggered your interest. To pursue this, apart from the further reading we have suggested, there are a number of good science dictionaries and the more reliable Internet sources that you can delve into.

As a final aid to your discovery, we thought it would be useful to illustrate some units of measure that might stretch your

Introduction

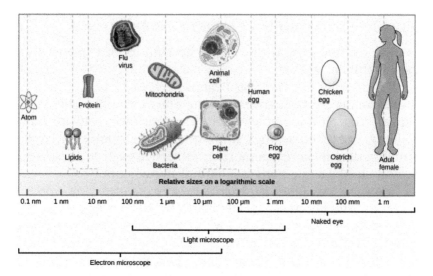

Figure 1.2 Diagram showing the range of dimensions on a logarithmic scale with illustrative examples of structures at a given size and the limits of visibility with the naked eye and various microscopes. A human hair is about 100µm wide – a similar size to a human egg. Emulsion droplets in skin-care lotion are about 1 µm wide – similar to bacterial cells.
Reproduced from https://commons.wikimedia.org/w/index.php?curid=49923763, under the terms of a CC BY 4.0 license, https://creativecommons.org/licenses/by/4.0/deed.en.

everyday experience of size. Figure 1.2 shows the different scales of measure of structures you will come across in this book – everything from lipid molecules to the full organism.

In this book, chapters have been written by an experienced authority on a topic, and in many cases they have paired up with someone who has recently started out on their career. This is intended to reflect their own journey of discovery and to help bring out some of the common questions that you, as a 'novice' reader, may also want to ask.

For those of you interested in finding out what a cosmetic scientist can end up doing, Box 1.2 contains a short interview with one of our own 'novice' authors.

Appendix 1 of this book provides a broader description of career opportunities. Whatever their scientific background, a cosmetic scientist will also have a passion for science and how it

BOX 1.2 WHAT DOES A COSMETIC SCIENTIST DO?

Here's an interview with one of the authors of Chapter 10, Rachael Polowyj.

What Is Your Job Title and What Do You Do?

I'm an Account Manager at an ingredient supplier where we source cosmetic ingredients from around the world for our clients. Our clients range from small indie brands to large multinationals, so it's important to service each of these accounts to the same high standard of ingredients.

What Are the Main Characteristics of Your Day-to-Day Work?

My day-to-day work usually starts by answering emails to new and existing clients. Mostly, brands contact me looking to create a product and they want to know how to make it. I will use my past formulation experience to suggest ingredients. I'll also work with our laboratories around the world to create some formulation samples for our clients. They can then try out our materials in a finished product before purchasing them.

What Is the Best Bit of Your Job?

One of my favourite parts of the role is that I get a first sight of the latest ingredients. Not only is that exciting in itself, but I enjoy the in-depth training on various scientific breakthroughs from the scientists who invent the novel ingredients. I then travel all over the country meeting with brands and manufacturers, teaching them about our innovative ingredients. It's quite rewarding when a customer selects your materials and goes on to claim them as 'hero' ingredients across their packaging. Another part I enjoy is working with my favourite brands and receiving free samples when their products are launched!

How Do You Use Your Scientific Training in Your Job?

As a trained cosmetic chemist, I get to use my formulation knowledge every day. I look after approximately 100 cosmetic companies so there are questions every day. The main queries involve practical advice on how to make formulas, but there can be many others. Why is my formula unstable? How can I make my self-tan last longer? How do I make my shower gel milder? – the list goes on! I would say within formulation, I use surface chemistry quite frequently to understand how surfactants are affecting surface tension between two substances. This is commonly where stability issues arise in my customers' formulations, causing them to separate.

What Other Departments/Professions Do You Work With?

I mainly work with research and development chemists, but also with marketing departments from brands. I find that marketing teams are increasingly important to work with as they are the team who communicate the ingredients to the end consumer. It is vital to explain the science behind the ingredient fully so that brands create a clear message for their consumers; there are so many horror stories in the media that are so misleading and untrue. Linked to this, I also work with journalists to discuss ingredients from a scientist's perspective.

How Would You Summarize Your Career Path So Far?

Now you ask – unbelievable! I graduated only three years ago, then started as a chemist in the laboratory and now somehow I've managed to land myself a job at management level! I just find it astonishing that massive brands ask me for advice and I completely love what I do. It's been a complete whirlwind – I've had my Masters research project published in one of the largest scientific magazines in our industry. I've won awards from the CTPA, the SCS Laura Marshall Award for the most innovative cosmetic product. I've had my research on display at the International Federation of Cosmetic Chemists in Germany. I have my own column in a global magazine where I review finished products on the market, and now of course I'm helping to write a chapter in this book! All in three years – I told you it was unbelievable!

Are There Any Specific Skills Essential to the Job You Do?

Undoubtedly, a passion for science and cosmetics is key. It's also useful to be creative and experimental as practical laboratory experience helps. I believe that trust is a very important thing in business, not just in the science but in building professional relationships. Confidence is equally important – in the science and understanding of your role. This comes in handy when you have to present new ingredients to a room full of experienced chemists or buyers.

> *What Qualifications Are Needed to Do Your Job? Which of Your Qualifications Do You Find Useful in Your Job?*
>
> A degree is important, preferably in a scientific subject. I have a Masters degree in Cosmetic Science, which covers a variety of sciences such as chemistry, microbiology, physics and biochemistry. I find that having an understanding of so many sciences is extremely helpful in my career. I use maths quite often for negotiating pricing with clients, so a general knowledge is fine.
>
> *Is There Anything You Wished You'd Studied When Younger That Would Be Useful Now?*
>
> I wish that I learned more languages at a higher level. I travel around Europe every few months meeting suppliers and, although I am very lucky that most people speak English, it has made me lazy and I've forgotten the basic language skills I once knew!

contributes to products. The authors of the chapters in this book pass on some of this in their writing.

Happy reading.
The Editors.

REFERENCE

1. S. E. Kemp, T. Hollowood and J. Hort, *Sensory Evaluation: A Practical Handbook*, Wiley-Blackwell, 2011.

CHAPTER 2

Clean Chemistry – The Science Behind Cleansing Products

EDWARD ROLLS

Symrise, UK
Email: edward.rolls@symrise.com

2.1 INTRODUCTION TO SURFACTANT BEHAVIOUR

I can honestly state that I find surfactants a truly interesting topic for cosmetic science. They are the key ingredients of many core products and, without them, formulating products such as shampoos, body washes, conditioners and many traditional emulsions would be very difficult. As is often the case with science, there is a lot of 'jargon' associated with this class of molecules, with many different terms and names, so I will attempt to describe all of this in the simplest terms and show where the different names come from and put them all into context.

2.1.1 So, What Makes Surfactants So Special?

We should start with the structure; the simplest way to start is to think of them as a tadpole as illustrated in Figure 2.1. Now, at

---

Discovering Cosmetic Science
Edited by Stephen Barton, Allan Eastham, Amanda Isom,
Denise McLaverty and Yi Ling Soong
© The Royal Society of Chemistry 2021
Published by the Royal Society of Chemistry, www.rsc.org

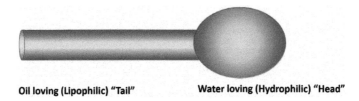

Figure 2.1 Basic surfactant structure.

first glance you may think, what is so extraordinary about this? Indeed, as a simplistic structure this means that there are only two distinct regions, a head and a tail. This, however, is the key to the characteristics of a surfactant molecule. If we start with the head, this is the portion that is water loving (*i.e.* it wants to be in water); it is therefore water soluble and does not want to be anywhere near oil. For a more scientific definition, the 'head' part of the surfactant could be described as follows:

- polar (likes to be in water – see the next few paragraphs for a description of polar);
- hydrophilic (literally water loving) or conversely lipophobic (literally oil hating).

Now, if we return to the 'tail' part of the molecule, this is completely the opposite – it is oil loving, so it can be described as follows:

- non-polar;
- lipophilic (oil loving) or conversely hydrophobic (water hating).

2.1.2 How Does the Special Structure Affect How Surfactants Behave?

As you can imagine, with there being so many terms to describe a surfactant it is easy to become confused with terminology, so to keep things simple I will just keep describing the different surfactant molecule portions as 'water-loving' and 'oil-loving' parts. Now, what is the relevance of all of this (if at this point you are struggling to share my enthusiasm!)? So, now imagine that we add some surfactant to water, what will happen? Well, the

Clean Chemistry – The Science Behind Cleansing Products 21

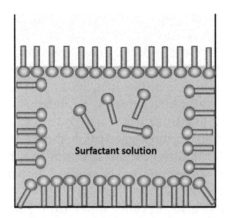

Figure 2.2 Surfactant solution. Surfactant molecules adsorbed at the air/water interface, the hydrophobic tails protrude from the surface of the water. Surfactant molecules adsorbed at the glass/water interface, the molecules are packed tightly together.

water-loving part of the molecule will be very happy as this is exactly where it wants to be. What about the oil-loving part, though? This does not want to be in water at all, so it will do everything it can to get out of the water. The surfactant molecule has two different parts that want to behave completely differently – this is called amphiphilic behaviour and the result is that the oil-loving part will pull the surfactant to the surface (called the surface interface) of the water. This is represented in Figure 2.2; around a beaker this is called a liquid/solid interface (where liquid and solid meet) and at the top it is called a liquid/air interface (where liquid and air meet). Now you will see where the name surfactant has come from: literally *surf*ace-*act*ive *agent*.

2.1.3 Why Isn't Using Water Alone Enough to Clean Things?

The structure that the surfactants form in water is the key for one of the first properties they offer that make them so useful in any cleansing formulation. Most of us will be aware that the water molecule consists of two hydrogen atoms and one oxygen atom (hence the well-known formula H_2O). The two hydrogen atoms have a positive charge and the oxygen atom is negatively charged. When they combine together in a water molecule the two hydrogen molecules place themselves just below (or above, depending on which way around you want to look) the oxygen atom. This creates an angle of 104.5° between the two hydrogens. The end where the two hydrogen atoms are has a positive charge and the end near the single oxygen atom is negative, meaning that the molecule is 'polarized'. This gives water all of its wonderful characteristics, leading it to be known as the 'universal solvent' – it dissolves many things, expands when frozen rather than contracting like most other materials and enables life, *etc.* We do not need to go into any further detail here, but one key effect of the polar nature of water is that it enables it to undergo surface tension, something anyone who has performed a belly flop will be aware of!

Surface tension also prevents water from spreading out over surfaces, which means that its ability to clean away (water-soluble) dirt is hindered. When surfactant is added to water as in Figure 2.2 the alignment of the surfactant molecules at the surface disrupts the surface tension, therefore reducing it. This is demonstrated in Figure 2.3a, where the droplet on the left depicts water on its own, with its high surface tension keeping its rounded droplet form. On the right surfactant has been added to the water, which reduces the surface tension and allows it to spread out, so it appears almost flat. This is also known as wetting out, hence surfactants are sometimes called wetting

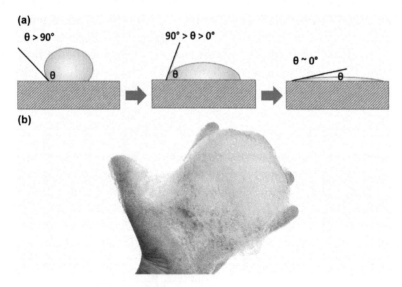

Figure 2.3 (a) Wetting behaviour; (b) soap bubbles (© Shutterstock).

agents. As you can see from the diagram, if you were trying to clean the surface in Figure 2.3 the water solution containing surfactant would do a far better job than the water alone because it covers a greater area. Another way of describing this phenomenon is that 'detergency' is increased; therefore, surfactants are also sometimes referred to as detergents.

2.1.4 Experiment at Home

If you want to try an experiment at home, you can try and balance a needle on the surface of water. This is easier if you place a small amount of tissue under the needle to start with, then gently remove it so that the needle remains floating on the surface thanks to the water's surface tension. If you would like some help, there are many videos available on YouTube, such as at www.youtube.com/watch?v=Tu9o4JXulvo. For a twist on this demonstration, if you gently add a small amount of washing-up liquid (away from the needle) you will see the needle sink; this occurs because the surface tension is broken.

2.1.5 How Do Surfactants Create Foam?

Now, if we return to the property of surfactants to break down the surface tension of water, this also leads to another phenomenon that you will all be aware of – foam. Foam is basically a 'mixture' of immiscible materials (materials that do not mix) and therefore a type of emulsion. In the case of foam in shower products, it is a mixture of air bubbles in liquids. Another common example of a foam that is not so obvious is whipped cream: when you whip cream you are literally whipping air into it. If you try to incorporate air into water alone, you will find it very difficult because the surface tension inhibits the entry of air. With the addition of surfactant and the corresponding reduction in surface tension, air is easily allowed into the solution, permitting the formation of foam. Foaming is a key characteristic of detergent products such as shower gels and shampoos; greater amounts of foam and good-quality foam are seen as premium and luxurious. Therefore, the right surfactants are key to creating the right type of foam. It should be emphasized that foam has no relation to how well the shower gel will clean; however, consumer perception is that a good foam is key to the cleaning power of the product, so creating a good foam is very important. This again gives us another term that is sometimes used to describe surfactants – a foaming agent. Again, it should be noted that some foaming agents are not surfactants, so it is not the best terminology to use.

In order to understand how surfactants enable this, we first need to understand what happens when we start to add higher concentrations of surfactants to water. If you go back to Figure 2.2, you can see that this is a good arrangement for the surfactants, both parts of their opposing characteristics are satisfied. The water-loving part is in water and the oil-loving part is either in air or against the glass. What happens, though, if we continue to add more surfactant and there is no more room at the glass/liquid interface and the liquid/air interface of the beaker? At this point, the oil-loving parts of the extra surfactant molecules added have nowhere to go. To remedy this unsatisfactory condition, the surfactants arrange themselves in ball-like structures as shown in Figure 2.4. When the surfactants are arranged in this ball-like structure, all the oil-loving tails face in together and are shielded away from the water, so a satisfactory arrangement once again occurs. A structure like this is called a micelle and again it offers many interesting characteristics that

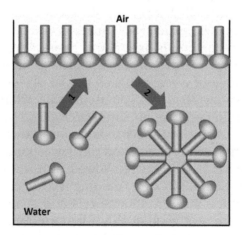

Figure 2.4 Micelle formation. (1) Initial aggregation to form an adsorbed monolayer on the surface; (2) surfactant molecules aggregate into micelles when the interface is full.

we can utilize for our cosmetic formulas. It should be noted that Figure 2.4 is a very simplistic representation of a micelle and that in reality they are of course 3D structures.

The point at which surfactant molecules form micelles rather than gathering at the interfaces is called the critical micelle concentration (CMC) of the surfactant. It is important that formulators know this to ensure they obtain the required properties from the surfactant solution.

If we go back to the problem of stabilizing foam, if you imagine an individual air bubble that has been incorporated into the solution, around the air bubble there will now be an extra interface where liquid and air meet. If you have added enough surfactant to surpass the CMC then the oil-loving part of the surfactant molecule will be attracted to the air bubble. A 'micelle' will form around the air bubble. In normal circumstances, any air bubbles in a liquid will quickly be attracted to each other and join together (coalesce) and escape. When surfactants are present, they form a barrier and stabilize the foam.

2.1.6 How Do Surfactants Help to Clean Dirt Away?

The final property that we will discuss here is solubilization. Put simply, this can be imagined by replacing an air bubble with surfactants surrounding a tiny oil droplet. There are two main benefits that we can utilize from this phenomenon. The first is detergency – we have already talked about how the wetting behaviour of surfactants adds to the detergency of surfactant-based liquids. Water, however, can only remove water-soluble dirt; when there is oil-soluble dirt present, the surfactant's oil-loving part will be attracted to the oil/dirt interface, and with some added agitation (such as with a sponge) the oil dirt can be broken up and will be 'solubilized' by the surfactant solution and can be washed away.

Finally, solubilization is a means to stabilize oil-soluble ingredients into a formula even when they are water based, the most common example being fragrance. The surfactant holds the oil droplet in place within the water (remember that oil and water normally do not mix). We will discuss the solubilization of fragrances and other oil-soluble ingredients in greater detail in a later section

SURFACTANTS

Just to summarize, a surfactant may be be referred to as the following:

- surfactant;
- surface-active agent;
- amphiphilic agent – it should be noted that not all amphiphilic agents are classical surfactants;
- wetting agent;
- detergent;
- tenside – this name has not been mentioned, but it is another term used to describe surfactants.

2.2 SURFACTANTS USED FOR CLEANING

So far, we have talked about the properties of surfactants and how they behave in certain conditions; now we will concentrate on surfactant types and focus on the most important ones for cleansing. Not all surfactants are the same and they can vary considerably in terms of structure and properties. We tend to classify surfactants into four main types (Figure 2.5): anionic, non-ionic, amphoteric and cationic.

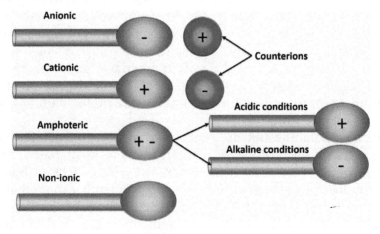

Figure 2.5 Different surfactant types.

Anionic surfactants are one of the most commonly used surfactants for cleansing in the cosmetic industry; they are characterized by having a water-loving head that has a negative electrical charge. Still one of the most commonly found examples of this is sodium laureth sulfate. Anionics (as we shall refer to them) tend to be the best foamers, create large volumes of foam and improve detergency. We have already mentioned sodium laureth sulfate (commonly abbreviated to SLES), which is one of the many 'sulfates' available to formulators. Before we discuss the different types, we should briefly look at their structure; rather than starting with SLES we will start with its close relative SLS (sodium lauryl sulfate, as seen on ingredients lists on products on the market), which is known as an alkyl sulfate.

DID YOU KNOW?

One of the hottest debates in the chemical industry is the spelling of sulphur and its derivative sulphate, or is it sulfur and sulfate? In 1990 the International Union of Pure and Applied Chemistry (IUPAC) formally adopted sulfur as the official spelling, but that seemed only to make the debate more intense. However, the IUPAC spelling sulfur is adopted throughout this book.

As we have already mentioned, a surfactant has both a water-loving and an oil-loving part; therefore, when producing a surfactant, we need two distinct types of starting material. For alkyl sulfates the oil-loving part comes from a class of ingredients called fatty alcohols, which then become the corresponding 'alkyl' part. The water-loving part comes from sulfate. Put simply, the fatty alcohol part is reacted with sulfur trioxide that forms an unstable ester (formed from an alcohol plus an acid), which is then stabilized by addition of an alkali [see eqn (2.1)].

$$C_{12}H_{25}OH + SO_3 + NaOH \rightarrow C_{12}H_{25}SO_4Na + H_2O \quad (2.1)$$

lauryl alcohol + sulfur trioxide + alkali
$\rightarrow$ sodium lauryl sulfate + water

Two things should be noted here:

- Structurally lauryl alcohol (Scheme 2.1) is a long chain, with carbon atoms at each intersection.
- Lauryl alcohol is derived from coconut and palm kernel and the carbon chain is never pure lauryl (C_{12}) – it will contain small amounts of capryl (C_{10}) and myristyl (C_{14}) chains.

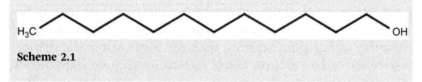

Scheme 2.1

For our SLS example, the fatty alcohol used is lauryl alcohol and the alkali used is sodium hydroxide (hence sodium lauryl sulfate). For alkyl sulfates, a range of fatty alcohols and alkalis can be used. Common examples of other alkalis that are used are ammonia, triethanolamine (TEA), monoethanolamine (MEA) and magnesium, and depending on the alkali used the surfactant product can have different properties. For example, the sodium class has the best thickening properties, the MEA and TEA types have the least irritant properties and sodium types can be the most irritant. This is summarized in Figure 2.6.

Alkyl sulfates were one of the first types of surfactant to be used; however, they have two main issues: they are relatively irritant (remember that for the last section of this chapter) and they are sensitive to hard water. To solve these problems, you can

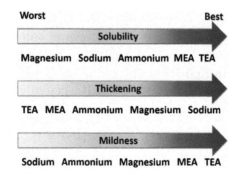

Figure 2.6 How the alkali affects alkyl sulfate performance.

use alkyl ether sulfates – the make-up of this class of surfactants is almost identical except that the starting fatty alcohol is ethoxylated, which means it is reacted with ethylene oxide (EO). You can react different amounts (moles) of EO, again giving various properties. The more EO added (usually not more than 3 or 4 moles) gives the mildest surfactant; however, when more EO is added it starts to reduce the quality of the foam and reduces the thickening properties. This is how we arrive at our previous example sodium laureth sulfate (SLES). Figure 2.7 summarizes the effects of different levels of EO.

As mentioned, SLES is still the most commonly used primary surfactant owing to its versatility. First, it is very cost-effective (cheap!); in such a competitive developed market, cost is always a key factor and usually the greatest constraint on what you can use in your formulation. In addition to being cheap it is very good at producing large amounts of foam. For more specialist products, you may consider using a milder surfactant such as MEA-laureth sulfate, such as in washes for sensitive skin and/or products for babies and children.

We have already mentioned that sulfates are excellent at cleaning, and the need for product quality leads us on to the final quality of SLES that makes it so commonly used – it is excellent at thickening. Again, thicker products are seen as more premium and luxurious. Before we go further here, we should briefly discuss 'thickness' and what it means. The scientific term and indeed the term used throughout the industry is 'viscosity'. Viscosity is the resistance to flow or deformation of a liquid which is caused by internal friction. If you think of milk, which

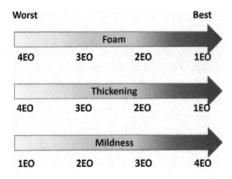

Figure 2.7 Effect of ethoxylation on alkyl ether sulfate performance.

behaves much like water, this has a low viscosity; golden syrup, on the other hand, flows much more slowly than milk and therefore it has a higher viscosity. From now on we will stick to the term viscosity.

In order to understand why and how SLES can be used to increase the viscosity of cleansing formulations such as bath foams and shower gels, we need to go back to micelles. We have already mentioned the CMC, which is the point at which there are too many surfactant molecules to all be able to align at the liquid/surface interfaces in a solution. At this point they start to form micelles. Now, if we imagine we continue to add SLES at higher and higher concentrations, not only will they start to form more and more micelles but they will also start to form larger ones. A system with a greater number of larger micelles will create more internal friction, causing resistance to flow (as mentioned earlier), and of course increase the viscosity of the solution.

If you think back to science lessons at school, you should remember that like charges repel (this is what is happening when magnets repel each other). We mentioned earlier that anionic surfactants such as SLES have a negative charge on their water-loving part. If you think of the (simplistic) micelle in Figure 2.4 earlier, you can see that if we packed more surfactant molecules into this structure the heads would approach closer and closer together, and at some point the negative charges that will start to repel each other will prevent the micelle from becoming any larger and therefore restrict any further increase in viscosity. One of the ways in which we can overcome this is to increase the ionic strength (a measure of the concentration of ions in the solution) of the water surrounding the micelle. If this happens, then the electrostatic repulsion (how much the ions are forced apart) between the surfactant heads is decreased, allowing the micelle to increase further in size, thus increasing the viscosity of the system. The simplest way to decrease the ionic strength of the water is to add ordinary table salt – sodium chloride (NaCl). I still believe that this is one of the most exciting observations for any newcomer to the industry – to see something as simple as salt transform a surfactant solution that has the viscosity of water into a viscous gel.

> **DID YOU KNOW?**
>
> Something as simple as salt can be used to thicken detergent-based products.

What invariably happens to chemists of all levels of experience is that at some point too much salt is added and the viscosity decreases. Micelles should not be thought of as static structures, they are constantly moving around, forming, breaking down and re-forming. If too much salt is added to the mixture, then the micelles become unstable and start to break down and re-form quicker. If more and more salt is added, there comes a point at which the life span of the micelle becomes shorter than the re-forming time, and at this point the viscosity decreases. If you were to measure the viscosity at regular points as you add more salt until the viscosity crashes, then you can plot a salt curve for your system. This can be an important exercise – when you develop products such as shower gels, they will have a viscosity specification to ensure that the viscosity is consistent from batch to batch. You do not want your viscosity specification to be near the top of the salt curve, otherwise during production it will be too easy to add to much salt and crash the viscosity, something that is not easily remedied.

So far we have looked at building a surfactant system just using SLES and salt. Although examples of very basic formulas can be found in the marketplace using only these two ingredients (plus the normal fragrance and preservative of course), it is better to utilize other types of surfactants to create a much more luxurious and workable system.

2.3 SECONDARY SURFACTANTS – LUXURIOUS, CREAMY FOAMS

If you were to buy a very basic cleanser (as mentioned in the previous section) and use it, you may notice a couple of things. Although it may produce a large amount of foam, the foam will just consist of large air bubbles. Although this is not necessarily a huge problem in terms of performance, a tighter foam that consists of small air bubbles will look and feel much more luxurious to the consumer. A foam that is luxurious is often described as being creamy. The other thing you may notice with the basic cleanser is that it may leave the skin feeling dry. Although we have already talked about how SLES is milder than SLS thanks to the addition of ethylene oxide, SLES on its own can be quite harsh to the skin and dry it out – something we refer to as 'stripping'. This happens because your skin has a lipid barrier to protect it and prevent moisture from escaping. When surfactants come into contact with the skin, they do not have the ability to distinguish between oily dirt and natural oils on your skin. Hence both can be removed, leaving a disrupted, incomplete barrier that allows moisture to escape – leaving the skin dry.

In addition to the stripping effect, surfactants can also be irritating to both the skin and eyes. Irritancy to the skin can lead to redness, swelling, itchiness and even stinging of the skin, and for the eyes the common symptom is stinging. Testing surfactants to grade how irritating they are to the skin is generally done using patch tests, in which the product is applied to the skin on a patch for a measured period of time, after which the skin is graded for redness, dryness and scaliness. Research has shown that irritation by surfactants is caused by the inhibition of skin proteins and enzymes that affect the skin barrier, causing it to break down further. The broken skin barrier then can lead to further penetration of surfactants and other irritating ingredients, such as preservatives and fragrances, which can escalate the problem further.

Eye irritancy is also tested and graded with a score. One of the simpler tests is the red blood cell (RBC) test. This test works by exposing surfactants to the blood and measuring how easily they cause the RBCs to split open (lyse). The ability to lyse the RBCs has been found to be correlated with how irritating the surfactants will be to the eyes. Another test is the HET-CAM test that uses an egg, which has a similar vascular tissue to an eye, to measure how the surfactant can damage the structure and give an irritancy score. Both of these tests are *in vitro* tests (they do not use animals).

It is generally accepted that for surfactants used in cleansers, anionic surfactants tend to be the most irritant, with amphoteric surfactants being less irritant and non-ionic surfactants the least irritant. It should be noted that different surfactants can act in different ways – some can be non-irritant to the eyes but can be 'stripping' when applied to the skin, and many other combinations. Less irritating surfactants are described as mild surfactants, which we will call them from now on.

As both amphoteric and non-ionic surfactants are milder than anionic surfactants, when they are mixed with anionic surfactants they have the ability to make the whole system milder. Both amphoteric and non-ionic surfactants are often called secondary surfactants as they are traditionally used at lower levels than anionic surfactants (which in turn are called primary surfactants). An amphoteric surfactant is classified as a surfactant that holds both a positive and a negative charge in its water-loving part. When held in a solution of water, if the pH is alkaline then the negative charge will dominate and if the pH is acidic then the positive charge will dominate. The most commonly found group of amphoterics in cosmetics is the betaines. Two main types are found, alkyl betaines, of which lauryl betaine is an example, and alkylamido betaines, of which cocamidopropyl betaine (which we will refer to as CAPB from now on) is an example. Although the betaines are the most commonly used amphoterics in cleansers, they are actually a poor example of an amphoteric surfactant as they do not gain an anionic nature at alkaline pH. A better example is the imidazoline types, of which the most commonly used is sodium cocoamphoacetate. If you look at the cleansing products you have at home, you may find some of these names in the list of ingredients.

At the time of writing, although trends are starting to change, most cleansers typically use SLES as the primary surfactant. The use level is anywhere between 7 and 12% in the product (and of course this is not fixed, it could be lower or higher). The most commonly used secondary surfactant is CAPB, often at a ratio 1 : 3 or 1 : 4 to SLES. As already mentioned, when used at these levels, the CAPB increases the mildness of your cleanser compared with using SLES alone, and it also helps benefit the foam. Whereas SLES will just produce large air bubbles (called polyhedron foam) that lack creaminess, CAPB creates smaller 'ball foam', which literally fills in the gaps and makes the foam tighter, denser and hence creamier.

The other useful quality of CAPB (and many other amphoteric or non-ionic surfactants) is how it helps with the viscosity of cleansers. We have already discussed how the SLES molecules formed larger and larger micelles, but the size of the micelle was limited by the negative charges repelling each other. When you add amphoteric (when they are within acidic conditions) or non-ionic surfactants, then these surfactant molecules can slot in between the anionic surfactant heads and allow the micelle to grow larger, as can be seen in Figure 2.8. Generally, when primary and secondary surfactants are combined to generate viscosity, rod-like micelles (which can also be described as sausage shaped) are formed. As you can imagine, a liquid with a high number of rod micelles that cannot easily slide past each other will be more viscous than a liquid with many spherical micelles that will flow around much more easily.

When using amphoterics, remember that the nature of the head depends on the pH of the solution, hence pH can have a considerable effect on the viscosity of the product. The shape of the micelles can also vary depending on the ratio of the size of the surfactant head to that of the tail. If the head is large in proportion to the tail, then only small spherical micelles can be formed, so these will not be as good as a surfactant with a larger tail in proportion to the head.

So, we now have a slightly more sophisticated system using a primary and a secondary surfactant that will be capable of higher viscosity, be milder for the skin and have a creamier foam. Now we are ready to apply the finishing touches, considered in the next section.

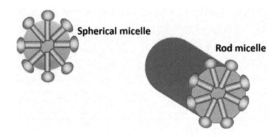

Figure 2.8 Micelle types.

2.4 FINISHING TOUCHES

So far, we have worked up to what could be almost a finished formula using water, SLES and CAPB; now we look at the finishing touches. If your product has a fragrance, it will need to be 'solubilized'. Solubilization is the final property that surfactants enable within liquids and again is very useful. The process of how solubilization works is very simple. If we think back to Figure 2.4, which showed how a micelle structure forms around an air bubble, then for solubilization of an oil the air bubble is replaced with an oil droplet. This solubilization allows oil-soluble ingredients to be added while still keeping the product transparent. In a classical emulsion such as milk, cream or, of course, a skin cream, which is basically oil droplets dispersed within water, the oil droplets diffract (bend) the light and do not allow it to travel through the emulsion, which is why it appears white. When a fragrance oil is adequately solubilized by surfactants, the structures are relatively small and, since usually only small amounts of fragrance are added, this allows the light still to travel through the system and hence keep it transparent. In order to keep the system transparent, it is important that the fragrance oil is solubilized in the correct way. In many surfactant systems with high levels of alkyl sulfate, sometimes the fragrance can just be stirred in at the end and the potency of the sulfates will solubilize it in. However, a more classical practice is to mix the fragrance oil with a suitable surfactant and then add the mixture to the cleanser after the primary and secondary surfactants have been added. Suitable surfactants tend to have a large head in proportion to the tail (the technical term is a high HLB, which stands for hydrophilic–lipophilic balance, which will be explained more in later chapters). A common solubilizer that is still used but is again succumbing to modern market trends is cocamide DEA (coconut oil diethanolamine condensate). What is useful about this ingredient is that it also further helps thickening by interacting with micelles and it stabilizes foam, making it creamier.

> **DID YOU KNOW?**
>
> Not all countries or regions of the world allow the same ingredients to be used in cosmetic and toiletry products. For example, the State of California has its own rules, which differ from those of the rest of the USA.

Because cocamide DEA is not allowed in the US State of California, many companies have looked at ways to replace it. Many other solubilizers are available, including ingredients such as PEG-40 hydrogenated castor oil and polysorbate 20. These can be added as part of the fragrance blend or added separately as a product is manufactured. They differ from cocamide DEA in that they tend to reduce the viscosity of the cleanser rather than help build viscosity. As you will see in the last section of this chapter, these ingredients are also not in favour with modern market trends and are being replaced with more natural alternatives that we discuss in the last section.

Solubilizing fragrances can be a tricky area, and a factor that can make it more difficult is the amount added. If a greater amount of fragrance is added there might not be enough surfactant molecules to take up all of the fragrance and the cleanser will lose its transparency. In this case, the fragrance may start to agglomerate (group together) and separate out from the cleanser. In this event, more solubilizer or a different type can be used to mix with the fragrance before addition to resolve the issue. Many companies offer solubilizer blends that can be more effective than just one alone; examples of these are mixtures of PEG-40 hydrogenated castor oil with trideceth-9 (PEG-9 tridecyl ether). High amounts of fragrance will obviously disrupt the micelle structure of the cleanser and therefore may reduce viscosity (I say 'may' because there are always exceptions where sometimes fragrances can help build viscosity). For these reasons, addition of more than 2% of fragrance to a surfactant, although not impossible, does become a real challenge.

All the ingredients that we have discussed so far are sufficient to make an adequate cleanser; we have discussed primary surfactants, secondary surfactants, fragrances and solubilizers. Beyond this, cleansers can be made more bespoke by the addition of actives and natural extracts. There is an endless list of potential extracts that can be used/sourced, in addition to the more familiar ones such as aloe vera and camomile. The easiest extracts to use are water-soluble products that can be mixed in with no effect other than adding colour to the formula (depending on the level added). Botanical oils can also be added but the same issues apply as with fragrances; they need to be solubilized in and if you already have a fragrance in your formula there might not be any extra room for more lipophilic material.

Behind adding botanicals, one of the key declarations that brands like to claim is 'moisturizing'. The term 'moisturizing' can be rather inaccurate when it comes to traditional cleansers. With the high level of surfactants (both primary and secondary), the reality is that most cleansers will trend to dry out the skin as they cleanse away the natural oils of the skin's lipid barrier. Even with the addition of moisturizers to a cleanser, the net effect is still likely to be that the skin will dry out – however, the skin will be less dry than if the cleanser was used without the moisturizer part added. This difference in how much the skin is dried out during use is what is used to claim a moisturizing effect. There are numerous ingredients that are used for this effect, including polyglyceryl esters such as polyglyceryl-4 caprate, glyceryl oleate, glyceryl laurate and other emollients that have PEGs added so that they can be added to the cleanser.

2.5 ALTERNATIVE SYSTEMS

Since I have been working in the cosmetics industry, momentum has been gathering for two trends, namely the 'free from' and 'natural ingredients' trends. It is often assumed that these two trends go hand in hand; however, it is my view that they should be thought of as two separate trends because, although you will find many natural brands that highlight ingredients that they are free from, just because an ingredient is natural does not mean that it is safer, milder or better than a synthetic alternative.

Part of these trends started with the desire for baby products that are very mild cleansers for newborns that avoid using harsh anionic surfactants. These trends have also gained traction because of the power of the Internet, which has led to consumers being able to gain more 'knowledge' about cosmetic products. I put 'knowledge' in inverted commas because the problem is that much of the information found on the Internet is misinformation or simply untrue. You will read more about myths in Chapter 10 and you can also visit the industry association (Cosmetic, Toiletry and Perfumery Association) website (www.ctpa.org.uk). I would recommend that everyone reading this book go and check it out (https://www.thefactsabout.co.uk).

> **DID YOU KNOW?**
>
> Sodium laureth sulfate (SLES) is the most common detergent used in shampoos and bubble baths.

You will read in Chapter 10 about the trends to move away from SLES and SLS. If a formulator wanted to avoid SLES in a formula, they could consider using a different sulfate, either not to use an ether sulfate or to select an alkyl sulfate with a milder base. A common example of this is ammonium lauryl sulfate (ALS).

Brands may choose to be completely 'sulfate free'; this will likely pose more of a problem. We discussed earlier that sulfates are used because they have many properties that are ideal for formulating cleansers, they foam well, build viscosity easily and are very cost-effective, and when you eliminate sulfates there is not one surfactant that you can use to replace the sulfate that will have all of these properties. One approach is to use isethionate-type surfactants, the most common of which is sodium lauroyl methyl isethionate. This is ideal in that it works in a very similar way to sulfates and can be used as a primary surfactant and can still be used with secondary surfactants such as CAPB. There is also a real benefit that this surfactant is milder than both SLS and SLES. The main differences compared with sulfates are that more care is needed to ensure a clear system and the standard grades require heating during manufacture (which adds time and extra energy consumption to the manufacturing process). Other newer grades are becoming available that come as a solution and do not need heating, so this group of surfactants really represents the future; however, price is still higher than that of sulfates.

Other alternative systems can also be used but each will come with its own challenges, *e.g.* sensitivity to pH, cost, effect on product viscosity and consumer acceptance.

2.6 NATURAL/ORGANIC AND SUSTAINABILITY

The final area to discuss is the rise of natural and organic standards and how these affect the formulation of surfactant systems. At the time of writing the three main options are COSMOS, NATRUE and ISO 16128. COSMOS and NATRUE are similar in that they heavily restrict what ingredients you can use in order to claim a natural or organic product. Both standards do not allow any alkyl ether sulfates (SLES, ALES, *etc.*). However, they do allow alkyl sulfates such as sodium lauryl sulfate, ammonium lauryl sulfate and the most popular choice sodium coco sulfate. Because of the chemical structure of this ingredient (the carbon chain length), it will contain some sodium lauryl sulfate (as an impurity), so even though SLS is not in the ingredient list the product should not be labelled as SLS free. As formulating for natural standards is largely driven by consumer perception, the use of surfactants with 'coco' rather than 'lauryl' in the name has a much better perception, hence the popularity of its use.

Now, if you think back to the second section of this chapter, you will remember that the alkyl sulfates have a greater skin irritation potential than the alkyl ether sulfates. Now that the naturalness of the alkyl sulfates is bringing them back into favour, care must be taken to ensure the use of suitable secondary surfactants to reduce the irritancy of the system. Although COSMOS currently allows CAPB, it is restricted as it has a significant synthetic non-natural content, hence the alkyl polyglucosides (APGs) and glutamate-based surfactants become very useful. APGs are very natural as they are made entirely from renewable sources, mainly sugar and either coconut or palm/palm kernel oil. Owing to their different chemical structures, there are different types that have differing properties such as mildness, thickening, foaming and solubilization. Choosing the correct surfactant will provide the desired product qualities.

> **DID YOU KNOW?**
>
> There is now an International Standard for the 'naturalness' of toiletry and cosmetic products: ISO 16128.

With ISO 16128, you can use whatever ingredients you like, it is simply an official standardized way to measure how natural the formula is. Considerations if you are using ISO 16128 is that ingredients such as CAPB and sodium cocoamphoacetate are more synthetic and bring the natural percentage down, whereas choosing alkyl sulfates will boost the natural percentage *versus* alkyl ether sulfates. Again, the glutamates and APGs are wholly derived from natural sources and so will boost the natural percentage. The ISO standard is self-regulating and has met with some controversy from the other natural/organic standards as they accuse it of confusing consumers as it allows all ingredients. In its defence, the ISO standard only allows ingredients that are already allowed by the EU – and if this is the case, they are deemed to be safe.

A final comment on sustainability – with the ever-increasing spotlight on the environment, sustainability is also becoming a major topic. For surfactants this is particularly relevant as many of them are derived from palm oil. There are increasing stories in the media about deforestation in Southeast Asia in order to plant more and more palm plantations, which not only impacts the environment with loss of rainforest but also affects the wildlife (especially highlighted with regard to orangutans). This area is a huge topic in itself, but there are a few ways in which the cosmetic industry is dealing with this issue that can be mentioned. One is to use alternative sources such as coconut and/or rapeseed; however, coconut trees cannot be grown in plantations, so the amount of oil is limited. Other oils such as rapeseed do not give anywhere near the same yield as palm oil – so an even larger area needs to be farmed to create the same amount of product, hence this is just shifting the problem. For now, use of RSPO (Roundtable on Sustainable Palm Oil) palm oil seems the best compromise. The idea is that the palm oil will come only from plantations where no new rainforest is being cleared. As a further compromise, there is also mass-balanced palm oil, which is a mixture of normal and RSPO palm oil. There is an RSPO website if you want to learn more about this (www.rspo.org).

We have covered a lot in this chapter and mentioned many types of surfactants. Table 2.1 shows some of the surfactants that you might commonly find in the products that you use and provides an indication of the properties that may have led to them being chosen for use in cosmetic products.

Table 2.1 Comparison of surfactants.

INCI[a] name	Chemical type	Surfactant class	Classical use	Foaming	Mildness
Sodium laureth sulfate	Alkyl ether sulfate	Anionic	Primary surfactant	+++++	++
Sodium coco sulfate	Alkyl sulfate	Anionic	Primary surfactant/natural organic	++++	++
Cocamidopropyl betaine	Amidoalkyl betaine	Amphoteric	Secondary surfactant	+++	+++
Sodium cocoamphoacetate	Imidazoline types	Amphoteric	Secondary surfactant	++++	+++
Sodium lauroyl sarcosinate	Amino acid based	Anionic	Alternative/sulfate free	++++	++++
Disodium cocoyl glutamate	Amino acid based	Anionic	Secondary surfactant/alternative	++++	++++
Disodium lauryl sulfosuccinate	Sulfosuccinate	Anionic	Alternative/sulfate free	+++	++
Sodium lauryl sulfoacetate	Alkyl sulfoacetate	Anionic	Alternative/sulfate free	+++	++
Sodium lauroyl methyl isethionate	Isethionate	Anionic	Sodium lauroyl methyl isethionate	++++	+++
Lauryl glucoside	Alkyl polyglucoside	Non-ionic	Secondary surfactant/natural organic formulas	+	+++++
Coco glucoside	Alkyl polyglucoside	Non-ionic	Secondary surfactant/natural organic formulas	+++	++++
Decyl glucoside	Alkyl polyglucoside	Non-ionic	Secondary surfactant/solubilizer/natural organic formulas	++++	++++
Capryl/caprylyl glucoside	Alkyl polyglucoside	Non-ionic	Secondary surfactant/solubilizer/natural organic formulas	++++	++++

[a]International Nomenclature of Cosmetic Ingredients.

2.7 CONCLUSION

The idea of this chapter is to serve as a brief introduction to surfactants and having read it, it is hoped that you will now understand the following:

- surfactant structure;
- how the structure of surfactants gives them their properties;
- types of surfactant: primary, secondary, anionic, amphoteric, *etc.*;
- basics of formulating a simple cleanser, including building viscosity and solubilizing fragrances;
- the challenges of new trends such as naturals and the use of milder surfactants.

There is, of course, no substitute for actually making products in the laboratory. When first starting out, cosmetics formulators spend a great deal of time experimenting with different amounts of surfactants and combinations of surfactants. Scientists will often have experienced the moment where the viscosity crashes: everything is a learning experience!

CHAPTER 3

Good Hair Day: The Science Behind Hair-care Products

P. CORNWELL*[a] AND J. LIM*[b]

[a] TRI Princeton, 601 Prospect Avenue, Princeton, NJ 08540, USA;
[b] Good Housekeeping Institute, London, UK
*Email: pcornwell@triprinceton.org; jasmine.lim@hearst.co.uk

3.1 INTRODUCTION

To most of us, our hair defines, to a large extent, how we present ourselves to the world around us (see Figure 3.1). Even in its absence, our scalp hair can define who we are and is our 'crowning glory'. As a result, hair care is seen by most consumers as one of the most important areas in cosmetics and personal care. Even in the face of great financial hardships, hair care often proves to be one of the last things people compromise on.

Hair has the ability to transform rapidly how people feel about themselves. Helping someone to have a 'good hair day' with a new hair style or look can instantly make them feel more positive and self-confident. Hair care is one of the most emotionally charged places to work in cosmetic science. As Coco Chanel once famously said, 'A woman who cuts her hair is about to change her life'.

Discovering Cosmetic Science
Edited by Stephen Barton, Allan Eastham, Amanda Isom,
Denise McLaverty and Yi Ling Soong
© The Royal Society of Chemistry 2021
Published by the Royal Society of Chemistry, www.rsc.org

Figure 3.1 Hair has been an important aspect of adornment for centuries and continues to be so. Left: *Girl with Cherries* by Ambrogio de Predis. Right: hair fashioned into box braids. Right image from https://commons.wikimedia.org/wiki/File:Hair.jpg under a CC BY-SA 2.0 license, https://creativecommons.org/licenses/by-sa/2.0/deed.en.

This chapter provides an overview of the science that underpins product development in the hair-care category. We start with the work that has been performed to understand the structure of hair and how hair growth occurs in the follicle. We also describe what can happen when hair 'goes wrong', for example, in hair damage, greying and hair loss. We subsequently cover the science that underpins hair-care product development, the chemistry that is used to create the products that we use and the measurement techniques used to test product performance and support claims.

3.2 HAIR STRUCTURE

So, what is hair? Like most simple questions, this one has a very long and complicated answer, but here we will summarize some of the brilliant science that has been done to give us some of the answers. Readers interested in the structural biology of hair can find more details in the book *The Hair Fibre: Proteins, Structure and Development*.[1]

BOX 3.1 THE EVOLUTIONARY BIOLOGY OF HAIR

Recent evolutionary biology work has shown that hair, scales and feathers probably all derive from the same embryonic structures across different animals. The evolutionary leap from scales to fur is difficult to define exactly, but probably occurred in the first common descendants of mammals, over 220 million years ago. Supporting this, fossils of an early mammal recently found in China, determined to be 125 million years old, clearly show fossilized fur (Figure 3.2).

Figure 3.2 Early Cretaceous mammal (*Eomaia scansoria*) from the Yixian Formation of Jehol Biota in China.
Reproduced from ref. 2, https://www.app.pan.pl/article/item/app51-393.html, under the terms of a CC BY 4.0 license, https://creativecommons.org/licenses/by/4.0/.

The evolution from early mammals to humans has meant that much of the thick body hair has been lost. Evolutionary biologists still argue about why we have lost our body hair. One plausible reason is that when our ancestors moved from the forests to hot grasslands, less fur and more sweat glands were needed to prevent overheating while hunting. Another reason might be that the loss of hair from the face may have been related to emotional communication. The reality is probably a combination of many reasons.

Anatomy students are taught that 'function always reflects structure'. It is the same with hair. Understanding hair structure helps us to understand why hair behaves as it does, leading to better designed treatments and products.

Most of us think of our 'hair' as the mass on our scalp, but there is much more to hair, as we shall see. Starting right from the beginning, hair is believed to have evolved from the same common ancestry as scales and feathers (see Box 3.1).

Hair, or fur, had many evolutionary advantages for these early mammals. It provided insulation, it helped to shed water and snow from the body quickly, it protected the skin from mechanical insults (scratching from burrowing or running through vegetation), it provided protection from the sun, it provided camouflage, it gave a heightened sense of touch and, importantly, it provided a means of signalling and attracting mates. It is often useful to remember these things when trying to understand the details of hair structure in humans, since these evolutionary advantages can help explain why things are built the way they are.

So, the result of this fascinating story is that humans have different hair types in different parts of the body: fine hairs over the face, legs, arms and abdomen – these are called vellus hairs – and thick hairs on the scalp, beard, eyebrows and eyelashes, under the arms and in the pubic area – these are called terminal hairs. Hair of intermediate thickness is also found on the arms and legs. Some more interesting hair facts are presented in Box 3.2.

All human hair has the same general structure (Figure 3.3). It consists of an inner core – the cortex – encased in a protective

---

BOX 3.2 HAIR FACTS YOU PROBABLY DIDN'T KNOW!

- There are approximately 120 000 hairs on a human head.
- There are 200–400 follicles per square centimetre of scalp area.
- Terminal hair on the scalp usually has a diameter of 60–100 µm.
- Each hair normally grows at about 1 cm per month, equivalent to a total of over 12 km of hair every year!
- Follicles in your scalp will produce roughly 1 g of new hair protein per day.

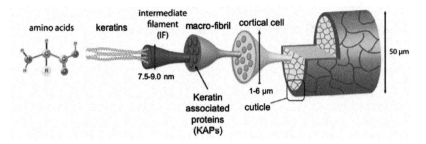

Figure 3.3 The structure of hair showing how the internal structures work together to produce a tough yet flexible fibre.
Adapted from ref. 3, https://doi.org/10.7717/peerj.619, under the terms of a CC BY 4.0 license, https://creativecommons.org/licenses/by/4.0/.

cuticle layer. The cortex makes up most of the hair mass and contains keratin proteins that provide much of hair's mechanical strength. Thicker hair may also have a medulla, which runs along the centre of the hair's length. The medulla consists of air-filled voids that provide thermal insulation.

The key structural proteins in the hair cortex are the keratins (see Box 3.3). These proteins provide much of the hair's tensile strength. Keratin proteins make up the strong rope-like structures in the cortex called intermediate filaments (IFs). Keratin proteins come in two main types, acidic type I keratins and neutral type II keratins. These two types of proteins coil tightly around each other in pairs, called protofilaments. Around 7–10 pairs of protofilaments then cluster together to form each IF (see Figure 3.3). Thousands of IFs are bunched together in each cortical cell.

At this stage of the story there is another interesting development. Proteins in the hair cortex are not just like a piece of rope, consisting of dry bundles of twisted IFs. In between the IFs are a different type of protein, imaginatively called keratin-associated proteins (KAPs). KAPs are globular-shaped proteins that are highly 'cross-linked' – these special types of chemical bonds form links with other KAPs and with IFs. This creates a flexible mesh-like structure that provides cushioning and flexibility (Figure 3.3). To take the rope analogy, they are like a soft rubber layer that separates the stiff, dry strings in the rope. In hair science, the concept of IFs surrounded by KAPs is known as the two-phase model of hair keratins and is used to describe

BOX 3.3 DISCOVERY OF THE ALPHA-HELIX

The uncovering of the molecular structure of hair keratins is, in fact, the story of the foundation of structural biology. The discovery of the alpha-helix in hair keratins by Linus Pauling in 1950 preceded the discovery of the structure of DNA, and helped him win a Nobel Prize in 1954 (Figure 3.4).

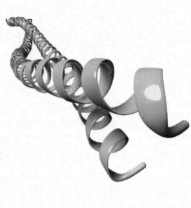

Figure 3.4 Left: Linus Pauling with 'rope'-like nature of the alpha-helix and (Smithsonian Institution/Science Photo Library). Right: a more up-to-date model representing the helical structures (Molekuul/Science Photo Library).

many of the mechanical properties of hair. The KAPs are vital to understanding the physical properties of the hair, since they absorb much more water than the IFs. As a result, hair swells much more width-wise than length-wise. In addition, absorption of water in the KAPs adds flexibility to the hair, and dramatically changes the hair's mechanical properties. High water levels reduce hair stiffness. The most obvious real-life example of this is hair frizz and loss of style when hair is exposed to high humidity. Absorption of water into the KAPs also makes the hair more fragile. You can test this on your own hair. Try stretching and breaking a few dry fibres and then try again when the hair been wetted for a few minutes. You will see how the wet hair stretches further and snaps much more easily. If hair is damaged in any way, it will snap even more easily.

The hair proteins in the cortex are packed inside spindle-shaped cortical cells. However, for everything to hold together, the cortical cells need to be tightly bound together by a structure known as the cell membrane complex (CMC). The CMC runs continuously between all the cells in the cortex and the cuticle and is the glue that holds everything together. Interestingly, the CMC between cortical cells is formed from two lipid bilayers (beta layers), sandwiching a layer of amorphous proteins (delta layer) (see Figure 3.5). You might ask, why doesn't hair have lots of rivet-like proteins joining all the cells together? Nobody really knows, but it is likely that the structure of the CMC gives enough flexibility between the cells to allow the hair to bend and twist without cracking in the joints. The CMC also has a role as a barrier to penetration of materials into the hair. It is, in addition, the main route of diffusion of low molecular weight materials

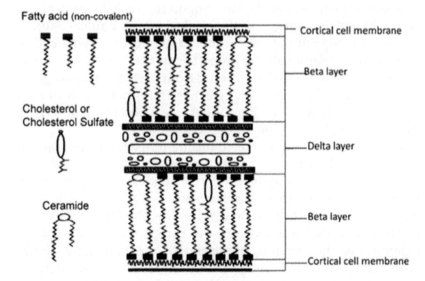

Figure 3.5 The cell membrane complex found between cortical cells: three related but different cellular cohesion components of mammalian hair fibres.
Reproduced from ref. 4, https://doi.org/10.1111/j.1468-2494.2010.00577_6.x, with permission from John Wiley and Sons, Copyright © 2010 The Authors. Journal compilation. © 2010 Society of Cosmetic Scientists and the Société Française de Cosmétologie.

into and out of the hair, rather like the intercellular lipids in the stratum corneum (see Section 5.1 in Chapter 5). In animals, the lipid bilayers in the CMC probably developed as a result of evolution to prevent the ingress of bacteria and moulds into fur – something very useful if your fur is often wet. Now, in humans, the CMC just irritates consumers by preventing the penetration of chemicals, including colourants, into the hair! For more on this, see Section 6.6 in Chapter 6.

So far we have spoken only about the structures in the hair cortex. The hair cuticle is also a very important part of the hair. It is approximately 5 μm thick and is comprised of 5–10 layers of cuticle cells. The overall structure of the cuticle has been described as being similar to the structure of a stack of paper cups, with each cup sitting inside another cup (see Figure 3.6). In hair, the walls of the cups, or the cuticle cells, are all angled slightly, with the lips of the cups, or scale edges, pointing away from the root ends. This is a deliberate design, as the reduced surface friction created going from root to tip helps with hair detangling. The stacked-cup design is also important to help control the effects of cuticle damage. With this design, mechanical damage to the scale edge of the cuticle should just reveal a fresh cuticle surface, maintaining all the properties of the hair. You can feel the difference in hair friction going from root to tip *versus* going from tip to root by simply taking a hair and running it both ways through your fingers. You may be surprised how much difference the cuticle scale direction makes to the friction.

The cuticle is the outer shield for the fibre, acting as a barrier to mechanical insults. It is also the wrapping that holds the cortical cells together tightly. Without a cuticle, cortical cells will tend to split apart, as you may have noticed if you get split ends. The cuticle, due to its location surrounding the fibre and its high mechanical stiffness relative to the cortex, plays a key role in determining the bending properties of the hair. As a result, attributes such as hair softness and movement are strongly influenced by the cuticle properties. Cuticle is damaged and removed by combing damage, sunlight damage and chemical treatments, such as bleaching. As a result, the number of cuticle layers often decreases from the root end of the hair to the tips. Complete removal of cuticle leads, as has been mentioned already, to split ends.

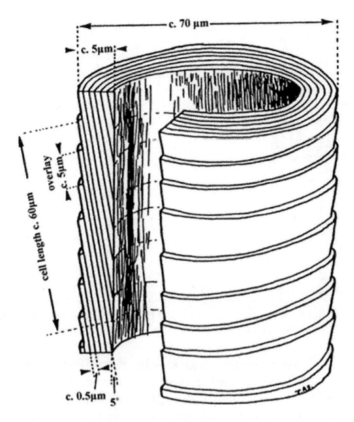

Figure 3.6 Hair cuticle. Stacked-cups model.
Reproduced from ref. 5 with permission from Society of Cosmetic Chemists, Copyright 1999.

The structure of cuticle cells is quite sophisticated (see Figure 3.7). Each cell is surrounded by a protein shell called the epicuticle. The epicuticle is, in turn, coated in a monolayer of covalently and non-covalently bound lipids. Having the lipids covalently attached means that they are permanently bound to the cuticle and cannot be washed away. On the parts of the cuticle exposed to the air, the lipid monolayer provides a water-repellent coating to the hair. The monolayer acts like the wax coating on a rain jacket, making the hair shower proof.

On the unexposed parts connecting cuticle cells, the lipid monolayer becomes part of the CMC holding the cells together and acting as a barrier to penetration into the hair. Inside their

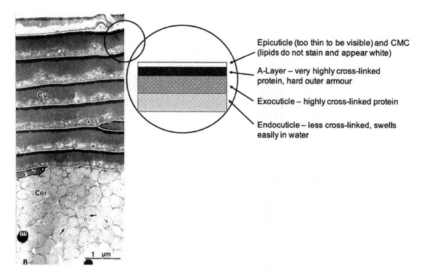

Figure 3.7 Structure of hair cuticle.
Adapted from ref. 6, http://dx.doi.org/10.4236/jcdsa.2013.32025, under the terms of a CC BY 4.0 license, https://creativecommons.org/licenses/by/4.0/.

protein shell, or epicuticle, cuticle cells comprise two zones – a very hard, 'exocuticle' and a more flexible 'endocuticle'. The difference in these zones is due to the degree of protein cross-linking. Cross-linking is a common process in biological chemistry and polymer chemistry. These cross-links create a more rigid and resistant protein structure, as seen, for example, in fibrin in blood clots. In the case of hair, the epicuticle is more highly cross-linked, creating a hard, outer protection for the cuticle. The endocuticle, being less cross-linked, is designed give the cuticle some flexibility.

In conclusion, the hair fibre is a very refined and sophisticated biomaterial, and its structure follows its many functions. We shall see in later sections how modern washing and styling practices can damage hair, and how modern hair products are tuned to work as effectively as possible on this complex material.

3.3 HAIR DIVERSITY

The previous section gave an introduction to hair structure and biology and described the key components of all types of hair.

However, just looking around us it is clear that there are a huge range of hair types. How is this the case? And what does this mean for hair product design?

The first thing you might notice about an individual is their hair colour. Hair colour is created by melanins in the cortex of the hair. Humans have just two types of melanin, pheomelanin, which produces red tones, and eumelanin, which produces brown–black tones. The rainbow of natural colours you see in hair is created just through the adjustment of the levels and ratios of these two melanins. Grey hair, as we will see later, is related to reductions in melanin in the emerging hair fibres.

The next big variation you might notice is the difference in curl. Human hair can go from being rod straight to being very curly. The reasons why hair can grow curly are quite complex. There are some differences in the way that the hair is formed in the follicle, and some differences in the structure of the cortical cells in the emerged hair fibres. Modern methods for the classification of hair curl now rely on precise measures of the curl patterns on sampled hair fibres. Hair can also vary in diameter. Figure 3.8 shows the variation in hair diameters for some different ethnic populations. There are two things to note here. First, Asian hair is often much thicker and coarser than Caucasian hair, whereas African hair, while being very elliptical in cross-section, is actually large in cross-sectional area. Second, the variation in hair diameter is very large even within an ethnic group. Some Caucasians have very fine hair and some have very thick hair. There is a huge overlap!

When designing a hair product range, it is always important to understand the hair types that you are targeting and how these affect hair habits, practices and needs. The challenge in hair care is that hair is so diverse both in its natural properties but also in its condition, a result of what people have done to it. It is not surprising that the hair section in the supermarket is so big! Some of the hair attributes that are commonly used in market segmentation include degree of curl, hair diameter (fine to coarse), hair density (sparse to abundant), hair length, level of hair greasiness, level of hair damage and susceptibility to frizz. Add to this the needs of some consumers who want to change their hair colour or hair curl, then you can see why this category is so diverse and challenging!

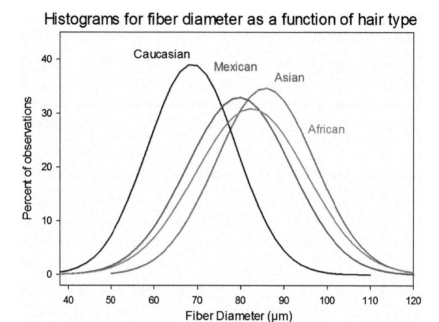

Figure 3.8 Hair diameter as a function of ethnic origin. Reproduced from ref. 7, https://doi.org/10.1111/ics.12509, under the terms of a CC BY 4.0 license, https://creativecommons.org/licenses/by/4.0/.

3.4 THE LIVING FOLLICLE

The previous section described the structure and function of hair fibres in some detail. We have seen that hair, after it emerges from the follicles, is essentially a 'dead' material. The cells that make up a hair fibre have lost their nuclei and are no longer able to function or divide as living cells. However, the biological processes that create hair in the first place are very much 'alive'. The relationship between the 'living follicle' and the 'dead' hair has long been a source of fascination for biologists and cosmetic scientists and is the route to solving problems such as hair thinning, hair loss and greying. This section introduces the key biological processes involved in hair growth. Later sections will cover hair loss and greying. Readers interested in learning more about follicle biology should look at detailed reviews by Schneider *et al.*[8] and Rogers.[9]

Each hair follicle on your head can be considered as a mini-organ. The hair follicle comprises the pilosebaceous unit (the part that the hair shaft grows from), together with associated structures such as the sebaceous gland, the apocrine (sweat) gland and the arrector pili muscle – the muscle that makes you hair 'stand on end' when you are cold or frightened. They have their own blood supply, nerve supply and immune systems.

The hair bulb, situated at the base of each hair follicle, is the 'factory' where new hair is made (Figure 3.9). Cells in the hair bulb are constantly dividing and differentiating to produce new hair. It is amazing to consider that the time needed for a new hair cell to be created and then transform into a fully formed, 'dead' hair cell can be just 48 h. The daily output from all your hair follicles is also astonishing. Given a normal growth rate of

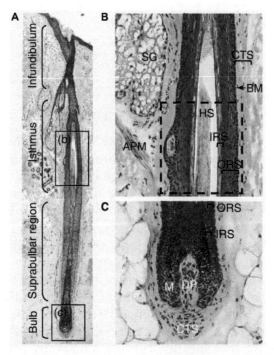

Figure 3.9 The human hair follicle in close-up. (A) The whole follicle with the hair bulb at the base and the infundibulum at the top where the follicle opens out onto the skin. Close-up image (B) shows the sebaceous gland (SG) and the arrector pili muscle (APM). Also shown are (from outside to inside) the connective tissue sheath (CTS), basement membrane (BM), outer root sheath (ORS), inner root sheath (IRS) and hair shaft (HS). The dashed squared area defines the bulge region. Close-up image (C) shows the main components of the hair bulb region with the dermal papilla (DP) and the matrix (M).
Reproduced from ref. 8 with permission from Elsevier, Copyright 2009.

0.4 mm per day, follicles in your scalp will produce roughly 1 g of new hair protein per day.

Hair growth doesn't just happen continuously. Hair follicles go through a growth cycle, which is divided into a growth phase, called anagen, a regression phase, called catagen, and a resting phase, called telogen. In humans, terminal hairs on the scalp have an anagen phase that usually lasts several years. Catagen usually lasts about 3 weeks and telogen lasts about 12 weeks. Hair can be lost in both catagen and telogen, although club hairs (hairs no longer growing but still anchored in the follicle) are usually lost in late telogen. This means that hair on your head is constantly being shed as some follicles move into the later stages of the hair cycle. As a result, most people will naturally shed approximately 100 hairs per day. For more on how the follicles keep producing new hair see Box 3.4.

BOX 3.4   STEM CELLS IN YOUR FOLLICLES

Stem cell biology is an area of great interest in science and medicine at present. Stem cells are cells with the unique ability to develop into a wide range of specialized cell types in the body. It is worth pointing out that stem cells are present in the follicle and are crucial to the hair cycle. Follicles, when they move into catagen, go through a process of killing off the cells in the bulb region, and the follicle shrinks towards the skin surface. To get back into anagen, stem cells in the higher regions of the follicle need to migrate to the bulb region and divide and differentiate to reform a new hair bulb. The stem cells then continue to divide to produce all of the different cells required to make the hair shaft and the root sheaths. This is a remarkable process and relies on stem cells. Biologists and cosmetic scientists are trying to understand better how these cell types function in order to see how they can be used for various treatments, including treatments for hair loss. It may be possible in the future to use stems cells to grow whole new follicles that can be regrafted into the scalp to treat hair loss.

The question is often asked, why do we have a hair cycle when it is such a complicated and demanding process? There are probably a number of reasons. The first reason is that having a hair cycle allows an organism to control hair length. If your hair just grew forever, it would grow to your feet! Also, shorter hairs, for example on the eyebrows, could not be made. A second reason is that in animals, where hair cycles are more coordinated over the body, fur can be shed and regrown in particular seasons of the year. Fortunately or not, depending on your view, humans do not shed all their hair in this seasonal fashion but there are small, subtle differences at different times of the year.

As we will see in later sections, the hair cycle can start to go wrong as you get older, leading to hair thinning and hair greying. Control of the hair cycle is also the route to many hair growth and anti-greying treatments.

3.5 SEBUM AND HAIR GREASINESS

Oily, greasy hair can be a problem for many people, especially those with straight, fine hair. Controlling greasy hair is the main reason why people use shampoos and other hair cleansing products. However, what most people don't appreciate is that the hair oils, or sebum, that give us 'greasy hair' have many useful functions on the scalp (see Box 3.5).

BOX 3.5   SEBUM AND THE SCALP – DID YOU KNOW?

- Sebum is produced by the sebaceous glands present in each hair follicle.
- A normal healthy scalp, in young adults, will produce 1.5–2.0 $mg\,cm^{-2}$ each day. Assuming a scalp surface area of 770 cm^2, then that is 1.2–1.5 g of sebum per day over the whole head.
- To put that amount of sebum into perspective, it is roughly equivalent to the 'pea-sized' amount of toothpaste you put on your toothbrush each morning.

Sebum plays many important roles on the skin, all over the body, not just the scalp, including skin surface lubrication and water-proofing. It has an essential role in controlling skin surface pH and affects the rate at which dead skin cells are lost from the skin surface. Sebum is believed to provide UV and free-radical protection. It also is absorbed into the skin and aids skin barrier formation.

It is also interesting to note that the high density of hair follicles and sebaceous glands, and thus the high levels of sebum, on the scalp make the scalp a unique breeding ground for specific types of oil-loving bacteria and yeasts. The scalp therefore has a unique skin microbiome. The follicles in healthy scalp skin, for example, are rich in bacteria called *Cutibacterium acnes* (which used to be called *Propionibacterium acnes* or *P. acnes* in older text books). These bacteria normally cause no problems, but when they grow out of control they can cause scalp acne. Scalp sebum also encourages the growth of a yeast called *Malassezia*. *Malassezia* is linked to conditions such as cradle cap, seborrheic dermatitis and dandruff.

So, how does sebum make your hair greasy? Well, sebum is present on the hair shaft as it emerges from the follicle, and it is also transferred from the skin to the hair when the two touch. Contrary to popular belief, sebum does not move from the scalp to the hair tips through any sort of capillary action – it moves only through direct transfer. High levels of sebum on the hair give rise to clumping and 'rats tails', reduced body and reduced shine. Experiments on hair tresses suggest that the point at which hair appears and feels greasy is affected by the degree of hair curl. Very curly hair is able to build up more sebum before looking and feeling greasy.

3.6 HAIR DAMAGE

So far we have seen how hair is built to reflect its many functions and how the hair follicle works to create new hair. Although a very carefully constructed material, consumers always want to look their best and so treat, wash and style their hair in many ways to make it look even more beautiful and fashionable. This is where hair can start to become damaged. Cosmetic scientists are always interested in how hair becomes damaged, as this opens up ideas for damage prevention and repair technologies (see Box 3.6). Readers interested in more detailed information about hair damage should consult the book *Chemical and Physical Behavior of Human Hair*.[10]

The key thing to understand about hair damage is that it is not just one process or mechanism. There are many different ways that hair fibres can be damaged.

BOX 3.6 COMMON QUESTIONS ABOUT HAIR DAMAGE

Why do I get split ends?

The outer layer of the hair – the cuticle – protects the inner cortex. As the hair grows, the free ends of the hair gradually become damaged. This can be from excessive repeated washing, or combing, or overheating from hair dryers or hair straighteners. This weakens the cuticle at the tips and regular friction leads to the complete removal of cuticle scales. Once the cortex is 'uncased' from the tight cuticle, the fibres in the cortex are free to expand and become splayed out, leading to split ends. In some cases, serious cuticle damage can occur higher up the hair shaft with a similar splitting effect; this may even result in breakage of the hair shaft.

Why does dark hair sometimes turn brown over time?

Hair browning can be caused by chemical changes, usually associated with the oxidation of proteins in the hair. This is mainly a result of heat styling but can also arise after UV exposure.

The most common way in which hair can become damaged is through washing and combing. This type of damage is sometimes called 'weathering'. The surfactants in hair products can extract hair proteins and hair lipids. Combing and rubbing actions and cyclic extension can also cause cuticle scale damage and crack formation. Weathering damage happens over months and years. This often leads to hair fibres having more damage at the tip end than the root end.

Heat styling with hair dryers, straightening irons and curling irons can also produce hair damage. Damage from heat styling can come from chemical changes inside the hair and from the physical effects of rapid water evaporation from the hair. Rapid water evaporation can create cuticle damage, such as cuticle bubbles and cracks.

Exposure to environmental extremes can also damage the hair. For example, UV radiation in sunlight can break the bonds between proteins, weaken the hair structure and lead to increased hair breakage. It can also damage melanin and cause hair lightening. UV exposure damages surface and CMC lipids, leading to altered surface friction and weakening of the cuticle. We are also learning more about the role of pollutant gases such as ozone and nitrogen oxides in damaging the hair. We are only just starting to understand these damage mechanisms, but it seems that chemical reactions involving free radicals may play a crucial role.

Hair colouring, using semi-permanent dyes, usually involves the use of peroxide treatments to lighten the hair (remove melanin) and to activate the colourant reactions (for more on this process see Chapter 6, Section 6.6). Peroxide treatments break protein bonds in the hair, increasing hair porosity and hair breakage. Peroxide also removes covalently bound lipids from the hair surface, creating a large change in surface properties. The hair surface becomes more water loving, or hydrophilic. This makes wet hair more prone to tangling and breakage, and also makes it harder for oily, or hydrophobic, materials to stick to the hair surface. As a result, silicone oil conditioning agents find it harder to spread on peroxide-damaged hair.

In conclusion, hair damage is not just one process. In reality, consumers' hair will be exposed to a combination of damaging treatments. Therefore, as we shall see later in this chapter, cosmetic scientists need to design hair products carefully to prevent and repair as much of this damage as possible.

3.7 HAIR THINNING AND HAIR LOSS

Varying degrees of hair thinning with age are well known and, unfortunately, come to most of us eventually. After a peak in middle age, hair fibres in both males and females become smaller in diameter with further ageing. In addition, the hair density on the scalp also decreases as the number of thick hairs growing from each follicle tends to reduce. Indeed, you could say that hair thinning is the hair equivalent of skin wrinkling.

It is linked to reduced hair body, volume and ease of styling. Consumers usually adapt to these changes by wearing shorter hair styles and by using volume-boosting products. Some consumers also use scalp massage to stimulate the scalp blood flow around the follicles, cosmetic scalp treatments and nutritional supplements.

This, the most common form of progressive hair loss, also commonly termed male pattern and female pattern baldness, is called androgenic alopecia. In the UK, it affects about 50% of men over the age of 50 and about 50% of women over the age of 65. It is dependent on genetic and hormonal factors and is linked to the effects of dihydrotestosterone, which is believed to cut off the blood supply to hair follicles. Hormones that contribute to 'male features' are called 'androgens', hence the name for the condition; it is important to understand that some androgenic hormones are also present in women. The exact reason why follicles in some parts of the body and not others are affected is still not fully understood. Cosmetic treatments include keratin fibre products to improve perceived hair density and root lift, alongside shampoos and conditioners for boosting hair volume. Cosmetic scalp treatments, with actives such as caffeine, have also been shown to have positive effects on hair growth, but do not completely reverse hair loss. Medicinal treatments for hair loss include minoxidil and finasteride.

Another important and serious hair loss condition is alopecia areata. This is usually related to localized, non-scarring hair loss, leaving bald patches. This type of hair loss is believed to be due to an autoimmune condition and is often associated with stress or emotional upheaval. Regrowth of the hair is possible but the condition can be extreme, with loss of hair affecting the whole body, resulting in complete hair loss. Unfortunately, there are no effective treatments for alopecia areata.

3.8 HAIR GREYING

The processes involved in creating hair colour and in greying are very complex. A detailed review was provided by Tobin in 2009.[11] Hair greying processes revolve around cells called melanocytes, which, as their name suggests, produce the melanins that make hair colour. Melanocytes are concentrated in the outer root sheath of an area close to the skin surface called the bulge region (see Figure 3.9). Why are they there? They are there so that when the follicle goes into catagen and the base of the old follicle is broken down and digested by the skin, they are able to remain safe and alive, higher up in the follicle. Then when, in anagen, the new hair bulb forms and the new hair begins to grow, some of the melanocytes migrate down to the bulb to start to produce melanin again. Hence the hair bulge region effectively contains a reservoir of melanocytes. As each hair follicle goes through several life cycles through your life, the numbers of melanocytes decrease until, at some point, the hair grows without any pigment, and turns grey/white. The rate at which your hair greys has genetic and oxidative stress influences. Unfortunately, there are not yet any cures for greying. Most consumers, if they are concerned, cover grey with hair colourants. However, cosmetic scalp treatments have recently been developed that claim to reduce the density of grey hairs.

Hair greying is often associated with hair thinning (see Section 3.7), as both are age related; however, when grey hairs first start to appear they can be coarser than the pigmented fibres. You might notice how, when greying first starts, the grey fibres sit out from the rest of your hair – very annoying!

3.9 SHAMPOO SURFACTANT BASES

As with most mammals, cleaning our hair is one of the most basic needs. As a result, for humans, shampoos are consistently the best-selling products in the hair category. Shampoos help to remove sebum and thus reduce hair greasiness. They also remove skin cells from the scalp, dirt particles and, of course, other hair products such as conditioning and styling agents.

People have used soaps of various kinds to clean their hair for many centuries. However, washing with soap was often avoided by women, as washing long hair with soap made it dry and tangled. In hard water areas, the soap would have also left waxy deposits on the hair.

Nowadays, people clean their hair with shampoos, dry shampoos, cleansing conditioners and shampoo bars. The habit of daily showering in developed countries means that people now wash their hair with shampoo on almost a daily basis.

Soaps and shampoos remove the sebum from the hair through the detergent effects of a key set of ingredients that they all contain, namely surfactants. As we saw in Chapter 2, one can visualize a typical surfactant molecule as looking rather like a tadpole, with a large, round head and a long, flexible tail.

In most cases, the head is water loving (or hydrophilic) and the tail is oil loving (or hydrophobic). To remind you of what you saw in Figures 2.2 and 2.4 in Chapter 2, this combination of water-loving and oil-loving properties within one molecule means that surfactants love to sit at the borders of things, for example, water and air, and water and oil. So, if you put a surfactant in water it will want to migrate to the border between the water and the air. The 'tadpole' structure allows it to sit with its head in the water and the tail pointing out to the air. Now, if you increase the concentration of your surfactant in water, as many tadpoles as possible will want to sit at the surface, but eventually the surface becomes full.

The next best thing for the tadpoles left in the water is to form clusters called micelles. Micelles allow the surfactants to arrange themselves with their hydrophobic tails all facing inwards and thus as far from the water as possible. Micelles are crucial to understanding how shampoos work. Shampoos contain high levels of surfactants that form micelles.

The insides of micelles contain all the oil-loving tail groups of the surfactants and so create a comfortable space for oily materials, such as sebum, to reside. Micelles explain how surfactants are such good oil detergents. Surfactants are classified by chemists according to the charge on their head groups; nonionics have no charge, anionics have a negative charge, cationics have a positive charge and amphoterics have both positive and negative charges. Anionics are usually the main type of surfactant used in shampoos and cationics are the main type used in conditioners. Amongst the most widely used anionic surfactants are sodium laureth sulfate and ammonium laureth sulfate.

The precise mechanisms through which surfactants in shampoos remove sebum from hair are not completely understood. A number of hypotheses exist, for example the roll-up hypothesis, which proposes that surfactants collect around the sebum on the hair surface and act to reduce the surface tension between the sebum and the water. This reduction in surface tension, it is suggested, helps the sebum to roll up into small oil droplets, become dislodged from the hair and be carried away. The micelle hypothesis suggests that individual sebum molecules diffuse from the hair surface into close-by surfactant micelles. The micelles then carry the sebum away. In reality, surfactants probably work through a combination of mechanisms.

Finally, surfactants in shampoos are not only used for their detergency effects. Modern shampoo formulations use blends of surfactants to create good foam properties, to build viscosity on the product and to enable hair conditioning polymers to work effectively. Not only that, but surfactants in shampoo formulas are carefully made to minimize skin and eye irritation. See Box 3.7 for some facts about greasy hair and shampoos.

Formulators, when they are developing new shampoo surfactant systems, use many different measurement techniques to test their prototype products. Analytical techniques, such as gas chromatography, are used to investigate the amounts and types of sebum oils removed. Specially designed foam testing equipment is used to measure the amount and type of foam generated. Many different types of devices are used to characterize the viscosity and flow properties of formulations. Cosmetic scientists also use a wide range of clinical tests and laboratory tests to understand the skin irritancy potential of the products. For more on this see Chapter 9, and for more details on shampoo surfactants see Cornwell.[12]

3.10 SHAMPOO CONDITIONING SYSTEMS

Modern two-in-one shampoo products, when you think about it, perform a rather fantastic magic act. They clean the hair at the same time as coating the hair with conditioning agents to help wet detangling and to make the hair feel smoother after it has been dried. How can this happen? How can both things occur at the same time?

> **BOX 3.7   FACTS ABOUT GREASY HAIR**
>
> *If I stopped washing my hair, would my sebum levels just balance out, leaving my hair naturally healthy and shiny but not greasy?*
>
> This is a very popular question. The short answer is 'probably not'. The very limited scientific evidence that exists suggests that sebum production levels in the scalp are not affected by wash frequency; the sebaceous glands just keep going regardless. However, more research is required here as some people do strongly believe that reducing washing frequency does control sebum output.
>
> *Did you know?*
>
> In the past, women often just combed their hair regularly ('100 combs a night') and then wore it tied up. Their hair probably stayed pretty greasy! This approach to 'grease control' redistributes the sebum – grease – throughout the hair; in addition to leaving the brush in need of a clean. It wasn't until the mid-twentieth century, and the mass production of synthetic surfactants, that shampoos, as we know them today, really took off.
>
> *Does my hair get used to my shampoo?*
>
> There's a popular belief that you should change your shampoo every now and again to keep your hair and scalp clean and in good condition. Although there is no precise answer, there are some scientific clues that might help guide you. First, studies have shown that different shampoo surfactant blends do remove different sebum components. Hence it is plausible that changing shampoos might be useful in removing all the different parts of the sebum from your hair. Second, shampoos deposit conditioning agents, such as cationic polymers, silicones, onto the hair and these can sometimes build up on the hair with repeated use. Hence it is possible that changing shampoos will help to control the build-up of conditioning agents.

The answer lies in the cationic conditioning polymers used in shampoos. These conditioning polymers were first introduced into shampoos in the 1970s and 1980s. They then gave rise to the first two-in-one shampoos, which combined cationic conditioning polymers and silicones.

Nowadays, nearly all shampoos contain cationic conditioning polymers and silicones, so there isn't really anything unique about being a two-in-one. Two-in-one shampoos rely on an aggregate of polymer molecules being formed between the cationic polymers and surfactants as the shampoo is diluted during the wash process. These aggregates, called coacervates, tend to stick to the surface of the hair and scalp. They give a good wet lubrication effect and aid with hair detangling. However, they can also be used to deposit other beneficial agents on the hair and scalp, such as oils, silicones and water-insoluble anti-dandruff agents. The most widely used cationic polymers are polyquaternium-10 and guar hydroxypropyltrimonium chloride. Commonly used silicones are polydimethylsiloxanes (dimethicones).

The conditioning benefits of cationic polymers are best felt during the wash process. After extensive rinsing, the coacervates break up and only a very thin layer of polymer is left on the hair. In order to achieve good dry conditioning effects, formulators tend to reach for lubricating oils and silicones. Once deposited and spread on the hair, these materials give added hair smoothness and ease of comb. We consider conditioning mechanisms in more detail in the next section.

Before leaving conditioning shampoos, it is worth mentioning build-up. The build-up of cationic polymers and silicones on the hair can be a problem if products are not correctly formulated. Good formulators will always test the effects of shampoos after repeated applications to ensure that conditioning benefits plateau at a suitable level and that build-up does not lead to greasiness and clumping.

3.11 HAIR CONDITIONERS

So, up-to-date surfactant-based shampoos clean the hair effectively and usually leave it in good condition. Why do we need conditioners? Well, to be totally honest, some people don't. If you have straight, undamaged hair and you cut it fairly short, then you could probably live comfortably without a conditioner. However, if you have curly hair, coloured hair, very damaged hair or long hair, then you usually need help with detangling your hair after you have washed it, to get it ready for styling. You might also, if you have long hair, need a conditioner to control hair static and flyaway.

So, how do conditioners actually work? The science of friction, wear and lubrication is called tribology. Tribology teaches us that there are two key types of lubrication effects on a rough surface such as hair. 'Boundary lubrication' means reducing friction using *thin*, broken films on the surface. Boundary lubricants have the effect of smoothing the surface by filling in gaps and crevices. 'Hydrodynamic lubrication' means reducing friction using a *thick* film of material on a rough surface. In this case, the surface coating submerges the bumps on the surface and your fingers or comb just glide over the surface.

Conditioners are perfectly designed for the job of detangling the hair. They use hydrodynamic conditioning to help your hairs glide over each other and detangle with great ease. Once rinsed out they also provide high levels of boundary lubrication to keep the hair feeling smooth and soft.

The key ingredients in hair conditioners, as with shampoos, are surfactants. However, conditioners always use cationic rather than anionic surfactants, because they form loose bonds with the negatively charged hair surface. In conditioners, cationic surfactants are mixed with fatty alcohols and water to create a unique structure called a lamellar phase. Lamellar means thin plates, scales or layers, and the ingredients in a conditioner have this type of structure. The molecular tadpoles that we talked about earlier, instead of being arranged in micelles, as in shampoos, are arranged side-by-side in flat layers, called lamellae. The lamellae in hair conditioners slide over each other when you spread the product on the hair and when you start to rinse it out. It is these structures that give conditioners the unique slippery feel that they have.

The conditioning performance of products is something that formulators take very seriously, as this is a key benefit for consumers. Combing tests on hair tresses are typically used to tweak formulations and optimize performance. Surface friction tests on tresses are also used to assess hair smoothness. Analytical tests can also be used to investigate the levels of cationic polymer and/or silicone deposition on the hair.

Now is perhaps a good time to mention flyaway. Hair flyaway is created by static building up on the hair. It is worst on very dry, winter days when the humidity is low, or in dry, air-conditioned offices. The lack of water on the surface of the hair means that charge cannot be dissipated, and static builds up between the hair fibres, pushing them apart. Flyaway is particularly bad on artificially coloured or very damaged hair, which has more surface charge than undamaged hair. The wonderful thing about the cationic surfactants used in conditioners is that they are able to neutralize the negative charge on the hair and control flyaway. Cationic polymers, oils and silicones in shampoos can do this job, but it is well known that conditioners are particularly good at controlling static and flyaway.

Conditioners usually contain oil or silicone conditioning agents to provide additional conditioning performance, particularly on dry hair. This raises many interesting questions for formulators. If you have oil or silicone in your shampoo, how much more do you need in your conditioner? And, are you putting oil or silicone on the hair from your two-in-one shampoo, only to wash most of it off when you apply your conditioner? Some formulators take the view that shampoos should be there just to clean, and that conditioners are there to condition, so they design shampoos formulated without any oil or silicone. Others argue that non-conditioning shampoos might be too harsh for most consumers and that one can never guarantee that consumers will actually use a conditioner. These formulators would formulate shampoos with reasonable levels of conditioning polymers, oils and silicones. There are no clear-cut answers to these questions. However, shampoos will affect the performance of a conditioner. It is always best practice to test conditioners with the relevant shampoo and, as mentioned before for shampoos, look at repeated applications to test for build-up.

As we discussed at the beginning of this chapter, hair is extremely diverse and market segmentation is always challenging. However, there are some things that formulators can do to adapt their shampoo and conditioners to suit different hair needs. For consumers with fine hair, looking for volume, then the trick is usually to dial-down conditioning levels. Consumers looking for volume usually need some help with getting some grip on the brush when blow-drying and when trying to achieve some root-lift, so lower conditioning is actually beneficial to them. High levels of silicones and oils also risk weighing down the hair, which is not what volume seekers want. At the other end of the scale, consumers looking to control and define their curly hair usually want higher than normal levels of conditioning. High levels of oils and silicones help with detangling curly hair while providing some cohesion between hair fibres to define the curl. Curly-hair consumers who use large quantities of styling products and leave-on conditioning products also benefit from having a deeper cleansing shampoo, which has the capacity to remove large amounts of oily product from the hair. Consumers with artificially coloured or damaged hair will have more problems with wet detangling. In addition, non-ionic silicones will find it difficult to adhere and spread on this type of hair. This type of hair benefits from higher levels of cationic polymers in the shampoo for better wet detangling, longer chain length cationic surfactants in the conditioner to provide heavier conditioning and cationic silicones for better deposition on damaged hair.

3.12 HAIR STYLING

Heat styling (blow-drying, use of straightening irons, *etc.*) works simply by driving the water out of your hair. Earlier in this chapter we learnt about the proteins in the hair cortex, the intermediate filaments and the KAPs. We saw how they form a two-component structure with the load-bearing filaments sitting inside a continuous phase of cross-linked KAP proteins that provide flexibility. When you heat style your hair, what you are doing is driving the water out of the KAPs and building more direct bonds between these proteins. As a result, the hair stiffens in the shape that you set it in. However, water will eventually penetrate back into the hair, break the inter-protein bonds and soften the structure again. What you then see is your style dropping out. Obviously, in wet or more humid conditions this happens faster, and hair frizzes quickly if exposed to rain and humidity.

Hair styling products, in general, are designed to reduce the effects of humidity as long as possible. They usually deposit polymers onto the hair that form cross-links between individual fibres, and which prevent the hair from moving out of style. The trick for formulators is to get enough polymer onto the hair to hold the style, while also giving the hair some natural movement, avoiding the 'helmet head'. This achieved by using polymers, and polymer blends, with just the right combination of adhesive properties, elasticity and stiffness. A substantial amount of work goes into getting this just right. Formulators typically use tress tests to investigate hold levels and the properties of the polymer films. A common test involves applying a styling polymer to a straight hair tress and then drying it. The tress is then placed across two pillars, to form a bridge. A device is then used to press the tress downwards in the middle, to test the bridge. Formulators acquire various types of information from a test like this, such as how stiff the styling polymer is, how easily the polymer cracks under pressure and how easily the styling polymer can re-form its hold after being 'cracked'. All of these properties help to predict how well the polymer blend will work in practice.

A good analogy for hair styling is the story of King Cnut (or Canute), the mediaeval Danish king who put his chair on the beach, sat down and tried to turn back the tide. Obviously he failed, and his feet got very wet! Well, the same thinking can be applied to hair styling. You can change the shape of your hair, going from, say, curly to straight by blow-drying or using a straightening iron, but, as in the story, whatever you do, the water will eventually come in and your hair will want to return to its natural style! The best that styling products can do is hold back the tide for as long as possible.

3.13 STRAIGHTENING TREATMENTS

For many years, people have tried to hold their hair permanently in style through chemical treatments (see Box 3.8). It wasn't until the early twentieth century that people started to experiment with chemically treating the hair on the head and creating the first 'perms'. In the present day there are, broadly, three types of permanent styling techniques, although each has many different

BOX 3.8 KING LOUIS XIV AND HIS WIGS

King Louis XIV of France was famous for bringing in the fashion for wearing wigs of long curly hair (Figure 3.10). Tresses of hair were rolled onto cylinders and boiled in water for 3 h and then dried in an oven. It is said that King Louis XIV had over 1000 wigs in his collection. As each wig was made from hair from up to 10 people, there must have been a lot of bald peasants in France in that period!

Figure 3.10 King Louis XIV of France.

variants. The first technique is permanent waving, commonly called a 'perm'. Perms can be used either to curl or to straighten the hair. This method relies on a reducing agent, a chemical that is able to donate electrons to other molecules and which is capable of breaking chemical bonds. Perms usually use a thioglycolate reducing agent that breaks some of the bonds that connect the KAPs to each other and the KAPs to the intermediate filaments. This allows the intermediate filaments to slide over each other into a new shape. A neutralization step then acts to re-form the bonds and set the hair in its new shape. Perms are not usually completely permanent and tend to drop out over time.

The second type of permanent styling technique is hair relaxation. This is used to straighten very curly hair and involves the use of high pH. Relaxation uses very strong reducing agents, such as sodium hydroxide. In these reactions, the bonds inside the hair between the KAPs are broken and a new set of permanent cross-links are formed. Relaxer treatments are damaging to the hair and are permanent.

The third type of permanent styling technique is hair straightening with formaldehyde, commonly known as a Brazilian keratin treatment. These use treatment with formaldehyde and heat to create new, permanent cross-links in the hair that hold it straight. Formaldehyde treatments are permanent. The issue with these treatments is that the release of formaldehyde from the hair during the heating stage creates a health risk to the customer and the hair stylist. Inhalation of formaldehyde gas can cause leukaemia and cancer. The use of treatments with high levels of formaldehyde is illegal in most markets, including the EU. Despite this, Brazilian keratin treatments are still used in some parts of the world. Such is the importance of having good hair! We would advise against this practice and safer alternatives are now also used, such as the use of glyoxylic acid.

When developing hair straightening or curling products, formulators tend to use a range of single fibre tests and hair tress tests to investigate the effectiveness of the treatments. The main aspect of interest is usually the rate at which the hair relaxes back into its old shape (if at all!). A straight hair fibre, for example, curled around a cylinder, should, after a chemical treatment, remain curled, even after dipping it in water. Water, as we know, destroys the simple water-wave, but should not affect a permanent chemical treatment. Formulators, when developing a perm, for example, will tweak the formulation ingredients to achieve the strongest and longest-lasting effect.

3.14 BRINGING IT ALL TOGETHER

We have seen how hair is literally our 'crowning glory' and how transforming our hair can have all sorts of effects on self-confidence and well-being. Our journey through hair science started by looking at how hair was created, through evolution, to be a very complex, multi-compartmented structure, perfectly adapted to its many roles. We also saw how diverse hair can be and what challenges that brings to cosmetic scientists and businesses.

Hair is, of course, created from the living hair follicle, and we have seen how the follicle works to produce new hair continuously, and how this can go wrong with ageing, greying and various hair conditions.

Along our journey, we also stopped off and looked at the role of sebum in scalp health and hair greasiness. We also looked at how washing, styling and hair treatment practices can damage our hair in different ways.

Armed with this knowledge, we then looked at how shampoos clean and condition our hair, and how scientists test products in the laboratory. We also looked at hair conditioners and how they work, and styling products and how they fight against your style falling apart. Finally, we took a look at how chemical treatments can permanently change hair styles, and how scientists approach testing these types of products.

Hair science is a very large and complex area, but it does make a great difference to people's lives. As the UK musician Morrisey once said, 'I do think that if your hair is wrong, your entire life is wrong'.

REFERENCES

1. J. Plowman, D. Harland and S. Deb-Choudhury, *The Hair Fibre: Proteins, Structure and Development*, Springer Singapore, 2018.
2. Z. Kielan-Jaworowska and J. H. Hurum, *Acta Palaeontol. Pol.*, 2006, **51**(3), 393–406.

3. F.-C. Yang, Y. Zhang and M. C. Rheinstädter, *PeerJ*, 2014, **2**, e619.
4. C. Robbins, *Int. J. Cosmet. Sci.*, 2010, **32**, 235.
5. J. A. Swift, *J. Cosmet. Sci.*, 1999, **50**(1), 23–47.
6. S. Sato, Y. Sasaki, A. Adachi and T. Omi, *J. Cosmet., Dermatol. Sci. Appl.*, 2013, **3**(2), 157–161.
7. S. Aslan, T. A. Evans, J. Wares, K. Norwood, Y. Idelcaid and D. Velkov, *Int. J. Cosmet. Sci.*, 2019, **41**, 36–45.
8. M. R. Schneider, R. Schmidt-Ullrich and R. Paus, *Curr. Biol.*, 2009, **19**(3), R132–R142.
9. G. E. Rogers, *Int. J. Dev. Biol.*, 2004, **48**, 163–170.
10. C. R. Robbins, *Chemical and Physical Behavior of Human Hair*, Springer-Verlag Berlin Heidelberg, 2012, ch. 5.
11. D. J. Tobin, *Int. J. Trichology*, 2009, 83–93.
12. P. A. Cornwell, *Int. J. Cosmet. Sci.*, 2018, **40**(1), 16–30.

CHAPTER 4

Oral Care – A Mouthful of Chemistry

ELEANOR ROBERTS*[a] AND STEPHEN MASON*[b]

[a] Beeline Science Communications, Ltd, London, UK;
[b] GSK Consumer Healthcare, Weybridge, Surrey, UK
*Email: beelinesciencecommunications@gmail.com;
stephen.x.mason@gsk.com

4.1 PHYSIOLOGY OF TEETH

4.1.1 Overview and Structure

As anyone who remembers the thrilling and slightly sick-making feel of a tooth first wobbling and then falling out as a child or adolescent, changes to the teeth can have a significant impact. Most people have two sets of teeth: 20 that develop in infancy, replaced by 32 permanent adult teeth, or up to four fewer in those who don't develop third molars ('wisdom teeth').

Permanent teeth function to cut (incisors), tear (canines or cupids), crush (premolars or bicuspids) or grind (molars). Each tooth is composed of the enamel crown, the visible part above the gumline, the neck, where the crown disappears into the

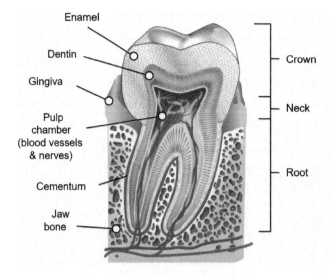

Figure 4.1 A human molar tooth.
Reproduced from https://commons.wikimedia.org/wiki/File:Human_tooth_diagram-en.svg from Wikimedia Commons by K. D. Schroeder, under a CC-BY-SA 4.0 license, https://creativecommons.org/licenses/by-sa/4.0/deed.en.

BOX 4.1   WHAT EXACTLY IS MY TOOTH COMPOSED OF?

- **Enamel:** The 2–3 mm thick, hard outer layer of the tooth.
- **Dentine:** Underlying the enamel, forming the bulk of the tooth from crown to root.
- **Pulp:** The inside chamber of the tooth, comprised of cells called odontoblasts, connective tissue, blood capillaries and the nerve network sounding like a faraway fantasy land: the plexus of Raschkow.
- **Cementum:** A thin, bone-like, connective tissue similar to dentine that covers the root of the tooth with a 0.1–0.5 mm layer; it attaches the tooth to periodontal ligament fibres in the jawbone and protects root dentine.
- **Gingiva (gum):** Soft tissue covering the jawbone, surrounding the enamel; the gingival margin is the thinnest part, where the gingiva meets the teeth.
- **Jaw bone:** The periodontal ligament of the jaw bone makes up the sockets (alveoli) that hold teeth in place.

gumline and enamel ends, and the root, the part below the gumline, composed of dentine, which anchors the tooth to the jawbone (Figure 4.1; see Box 4.1).

4.1.2 Enamel and Dentine

Enamel is 98% hydroxyapatite [$Ca_5(PO_4)_3OH$], formed from calcium and phosphate, and 2% water, some keratin (the same protein as in hair and nails) and a number of other trace substances such as zinc and chloride. The crystalline structure of enamel makes it the hardest material in the body and typically gives it a translucent, white appearance.

Like many minerals, although appearing very solid, hydroxyapatite can be weakened, in this case by a pH lower than 5.5. At this point, calcium and phosphate are lost from the tooth in a process known as demineralization. As soon as the pH rises, hydroxyapatite re-forms on the enamel, a process known as remineralization. Although at first it may seem shocking that the very top layer of enamel that we thought so solid can just dissolve, this cycle occurs naturally throughout the day with lost minerals being seeded by saliva.

Underlying the enamel, forming the bulk of the tooth, is dentine, which is 70% hydroxyapatite and 30% water, collagen (the protein found in skin) and trace impurities. Dentine is softer and more porous than enamel and, if exposed, is more prone to physical wear and decay. Dentine contains over 10 000 tubules per mm^2 that run from the pulp to the enamel (above the gum) or cementum (below the gum). Tubules contain a fluid and, at the base, protrusions of the odontoblast cells that make up part of the pulp.

> Dental enamel is so hard it can survive tragic circumstances such as when a body is damaged beyond recognition by fire or badly decomposed. As teeth, like fingerprints, are highly individual, a forensic dental expert can use dental records to match the remaining teeth to the person to which they belonged.

4.1.3 Saliva

Although you may think that saliva's only function is to lubricate and help digest food, it also has a protective

The Salivary Glands

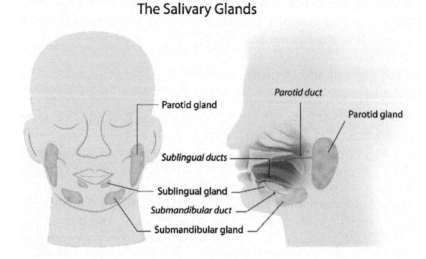

Figure 4.2 The salivary glands.

role, stopping oral tissues from drying out, housing antibodies and antibacterial factors and preserving a pH between 6.2 and 7.4. Ask anyone with xerostomia, or dry mouth, about the sometimes excruciating pain caused by lack of saliva and you will understand the importance of this vital fluid (Figure 4.2).

Saliva is not just frothy water. Although the main bulk of the tooth cannot be remade when lost (except in a laboratory), at the microscopic level the ions at the outermost layer of enamel are exchanged daily thanks to the calcium and phosphate found in this fluid. Additional salivary components include other ions (sodium, potassium, chloride, bicarbonate, fluoride), carbohydrates (fructose and sucrose), proteins (including those that form mucus) and enzymes (*e.g.* amylase, which breaks down starch), water-soluble vitamins (vitamin C, B vitamins), growth factors, hormones, antibodies, antimicrobial agents and immunoglobulins.

4.1.4 The Pellicle

Saliva also helps by forming the glycoprotein-rich pellicle, a 1–3 μm thick protective mucous layer covering the tooth surface.

This thin film is mostly composed of salivary protein, with some carbohydrates, lipid and inorganic residue along with trace amounts of calcium, phosphorus and fluoride. When left for a long time without the cleaning activities of twice-daily toothbrushing, bacteria can take up residence in the pellicle and dental plaque may form. The pellicle can also retain chromogens, the coloured part of food, beverages and cigarette smoke that cause the teeth to become discoloured.

4.2 WHEN GOOD MOUTHS GO BAD

4.2.1 Plaque (aka Dental Biofilm)

Plaque forms when oral bacteria are attracted to the pellicle. The initial bacterial colonizers of plaque are mostly aerobic Gram-positive species such as *Streptococci*. These feed on the dietary carbohydrates sugar and starch, resulting in production of acid, most notably lactic acid, and polysaccharide, the stickiness of which glues the plaque biofilm together. This initial plaque usually resides above the gumline and so is known as 'supragingival plaque'.

The polysaccharide element of this biofilm becomes the substrate for secondary colonizers, which include anaerobic Gram-positive and Gram-negative bacteria such as *Lactobacilli*, *Porphyromonas gingivalis*, *Actinomycetes* species and *Fusobacterium nucleatum*, alongside yeasts, viruses, protozoa and, from the host, antibodies and white blood cells. These feed on proteins from the oral cavity and produced by other plaque bacteria, resulting in the production of acids, amines, ammonia and volatile sulfur compounds. This plaque is more 'mature' and tends to form in less accessible places below the gumline, and hence is known as 'subgingival plaque'.

We all have plaque to some extent but it can be controlled by mechanical (tooth brushing), physical (abrasives in toothpaste) and chemical (components in toothpaste and mouthwash) means. Plaque tends to build up in areas less accessible to mechanical/abrasive removal, such as between the teeth, in enamel pits/fissures and at the gum margin. This is why dental professionals stress the importance of careful brushing across the four quadrants of the mouth for 2 min twice per day.

> Have you ever noticed that some power toothbrushes buzz differently every so often but aren't sure why? That's the 30 s time period you should spend on each quadrant in your mouth.

4.2.2 Dental Calculus (aka Tartar)

Calculus is the hard, white to dark brown crystalline deposit that occurs predominantly at the part of your teeth next to the gum and near salivary glands at the front of the lower teeth and back of the upper teeth. Calculus is mostly inorganic calcium phosphate, but also includes bacteria, lipids, carbohydrates and proteins. It has a much stronger attachment to teeth than plaque and is best removed by a dental professional and their tools as the appearance can be very similar to that of dentine.

4.2.3 Periodontal Disease (Gum Disease)

Periodontal disease is inflammation of the gingivae (gingivitis) or periodontium (periodontitis, involving the gingivae, connective tissue and underlying bone and ligaments). It affects around half of all people and becomes more common as a person ages. Periodontal disease can be brought about by plaque and calculus deposition and be due to both the physical irritation that these can cause and as a reaction to the bacteria and bacterial toxins housed within. Inflammation can also occur due to diet and lifestyle factors (such as smoking), genetic disposition and age, and sometimes because of over-scrubbing of gums with a hard toothbrush for too long or too vigorously.

> During a dental examination, your gums may be prodded with a sharp tool – this is testing to see if the gums bleed rapidly, an indication of periodontal disease. The examination may also include a probe with a smooth end being gently inserted between your teeth and gums – this is to assess if the gum has become detached due to gum disease.

Gingivitis is a precursor to periodontal disease and is inflammation next to the tooth. It can be reversed and limited by effective oral hygiene, including regular plaque removal. Signs of gingivitis include gums that are red, swollen and, sometimes, hotter than the rest of the mouth (Figure 4.3). Inflamed gums should be spotted by a dental professional but can also first be noticed by someone at home when they observe signs of bleeding occurring regularly following toothbrushing and/or flossing.

Periodontitis is the disease that occurs when gum and bone tissue break down. This is most often due to continued inflammation after gingivitis and is not reversible. The swelling that occurs in gingivitis leads to detachment of the gingiva from the tooth surface (Figure 4.3). More enamel and root dentine is exposed, which is colonized by more plaque, covered in more calculus, and the destruction continues. Periodontitis can continue through the gums to the underlying alveolar bone, which can itself degenerate, leading to tooth loss. Periodontitis is thought to affect around 15% of the population globally.

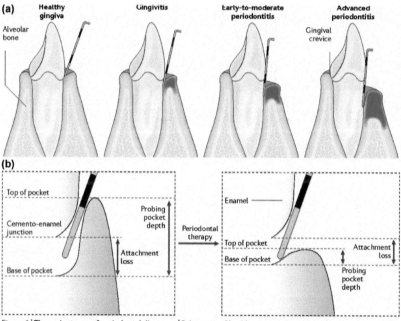

Figure 4.3 The main stages of periodontal disease.
Reproduced from ref. 1 with permission from Springer Nature, Copyright 2017.

4.2.4 Tooth Decay (aka Dental Caries)

The most prevalent major oral health disease in the world is dental caries. To this day, it still affects almost everyone on the planet at some point in their life and is considered the most common chronic disease in children. Dietary fermentable carbohydrates, such as sugars, are metabolized by plaque bacteria to produce acid. This is usually washed away quickly by saliva such that the plaque biofilm is at neutral pH most of the time, allowing any demineralized enamel to remineralize (Figure 4.4a).

Dental caries occur when the availability of fermentable carbohydrates is extended and/or frequent. This means that the pH of plaque falls below 5.5, when enamel demineralization occurs, for longer periods of time, which can mean that the remineralization process is inadequate and enamel (and sometimes dentine) is lost (Figure 4.4b).

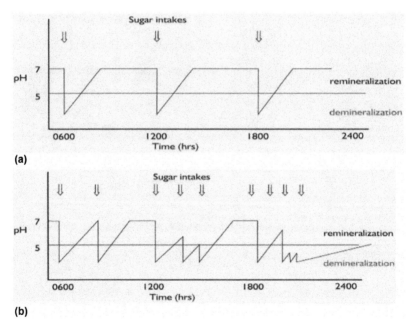

Figure 4.4 Effects of (a) infrequent sugar intakes and (b) frequent sugar intakes. Reproduced from ref. 2 with permission from Springer Nature, Copyright 2019.

If, at sometimes microscopic regions, the top layer of hydroxyapatite is not properly re-formed, leading to a softer enamel, the underlying, subsurface, hydroxyapatite, can also demineralize, causing pits of enamel loss that show as opaque/white spots known as 'precarious lesions'. A cavity quickly becomes filled with dental plaque bacteria and demineralization accelerates, leading eventually to a surface cavity.

> Oral bacteria aren't all bad. They help to maintain the protective biofilm above the enamel and, whereas some produce the lactic acid associated with dental caries, others use this as their own food source or produce alkaline by-products from metabolizing the salivary components urea and arginine (a protein).

As the look and feel of the white spots are different, they can usually be spotted by a dental professional using both probes and radiographs (X-rays). At early stages, careful use of toothpastes can repair the damage. This needs to be combined with adjustment of any dietary factors and lifestyle behaviours that led to the lesion in the first place.

However, what may appear as solid tooth could require a more extensive filling due to the underlying structure having decayed away. If demineralization continues, the enamel can be lost down to the much softer dentine. If the pulp is then exposed, the bacterial invasion of living tissue will lead to infection and a tooth abscess. Tooth extraction or significant (often expensive) dental restoration is then required.

For those who use fluoride toothpastes daily, the top layer of enamel becomes more composed of hydroxyfluorapatite (see Section 4.3.2 for details), a substance that does not demineralize until the pH is below 4.5. This helps to protect and prevent dental caries, which is why regular use of fluoride toothpastes twice per day, every day, is the dental public health strategy recommended by most dental associations globally.

As the natural 'window cleaning fluid' of the teeth, saliva is essential for washing away bacterial acid and for retaining a neutral pH. As such, dental caries may also occur at a higher rate in those with medical conditions or taking certain medications that reduce the amount of saliva in the mouth, potentially requiring the use of an artificial saliva product. Dental caries may also occur at a higher rate in those whose teeth are in a position where some parts are hard to reach with mechanical cleaning, including people wearing orthodontic appliances and partial dentures. Various components of oral care products (discussed later) can help bring plaque-controlling ingredients in contact with as much of a tooth as possible, beyond that reached by a toothbrush. Unfortunately, there are also people who naturally have thin enamel, and good dental hygiene is especially essential for those affected.

4.2.5 Tooth Wear (aka Dental Erosion)

Tooth wear is a condition that happens to us all to some extent throughout our life. It occurs *via* erosion, abrasion or attrition. Erosive tooth wear is, like tooth decay, due to a lower oral pH and dissolution of the enamel; however, it occurs at the tooth surface and does not involve the actions of bacteria. Erosion due to tooth wear is *via* direct dietary factors, such as acidic food and drinks, or internal factors, such as regurgitation of stomach acid (Figure 4.5).

The incidence of erosive tooth wear is rising, possibly due to the increased consumption of fizzy drinks, including mineral water, acidic 'sports' drinks and fruit juices. Erosive tooth wear is

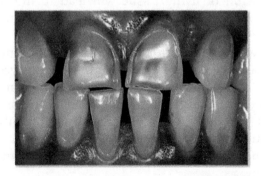

Figure 4.5 Severe facial erosive tooth wear in a 25-year-old, due to drinking fruit juice and holding lemon slices under the lip.
Reproduced from ref. 3 with permission from Karger Publishers, Copyright 2006.

more difficult to spot than tooth decay as it appears as a slow loss of enamel, leading to a smoother surface that can be either more shiny or duller than usual. On the side of the tooth, erosive tooth wear can occur as small, rounded cups.

> Saliva is usually all that's needed to dilute and wash away the acid from food and drinks, but if consumption is extended (*e.g.*, sipping mineral water or fruit juice regularly throughout the day, slowly eating an orange, sucking an acidic sweet), the pH of the saliva may not be sufficient to increase the oral pH over that needed to keep enamel intact.

Abrasive tooth wear is due to excessive tooth brushing: too long, too vigorous and/or too hard. It can also occur when using toothpastes or powders that are highly abrasive. Tooth wear due to abrasion most often occurs at the gingival margin where the tooth meets the gum. The enamel is thinner here and gums are prone to inflammation. Attrition is caused by excessive tooth grinding or clenching or the action of mouth jewellery such as a tongue or cheek piercing.

Erosion, abrasion and attrition, although described separately, occur together, not independently. This means that damage can be multifactorial, for instance through brushing (and especially over-scrubbing) of teeth soon after exposure to an acidic drink or vomiting.

4.2.6 Dentinal Hypersensitivity (aka 'Sensitive Teeth')

Sensitivity is a condition that causes a short, sharp pain when the tooth is exposed to cold and hot foods/beverages and other stimuli. Enamel erosion and receding gums can expose dentine both above and below the gumline. If this occurs, dentine tubules lose their usual sealed top, which opens a passage from the outside down to the odontoblasts, which extend from the pulp into the tubule (Figure 4.6). Dentine tubules are full of liquid that can be moved by various means when exposed: temperature differences (hot or cold food, beverages or air); pH changes (acidic food/beverages or dental treatments); tactile pressure (touch from a toothbrush or dental instrument) or osmotic

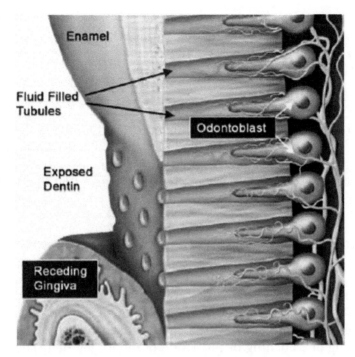

Figure 4.6 Exposed dentine tubules due to gum recession can lead to stimulation of dentine-associated nerve endings.

pressure (differences in fluid concentration caused by, most often, sugars). This movement is reflected in movement in the odontoblast, which results in a pain signal to the nerve endings in the pulp.

Dentinal hypersensitivity may be due to a lack of dental hygiene but also to enamel loss/gum recession following over-vigorous use of a toothbrush, especially in a scrubbing as opposed to a circular manner, or to brushing too often. Dentinal hypersensitivity does not occur in all teeth where dentine is exposed and may recede naturally. This can be due to the formation of a dentine 'smear layer', which is similar to the enamel pellicle but includes calcium and, when used, some toothpaste ingredients. Although protective, this layer can be removed by acidic drinks and fruits, alcohol and some herbal teas, and also by over-vigorous brushing.

Dentinal hypersensitivity can be counteracted by good dental hygiene, the use of over-the-counter toothpastes/mouthwashes that have hypersensitivity-targeting ingredients (see Section 4.3.4) or more specialist treatments from a dental professional.

4.2.7 Tooth Stain and Whitening

Tooth stain is a cosmetic imperfection predominantly on the outside of enamel (see Box 4.2). Natural tooth colour is mostly determined by how thick and translucent the white enamel is over the yellower dentine. External tooth staining is actually staining of the pellicle, plaque and calculus. It occurs at a higher level on rougher enamel, irregular teeth, when oral hygiene is poor and when consumption of staining factors is high. External tooth staining may be controlled by the action of regular

---

BOX 4.2   WHAT CAUSES TOOTH STAIN?

External
- Foods/drinks containing chromogens including tea, coffee, turmeric, red wine.
- Tobacco smoke.
- Complexing of the pellicle with metal ions such as tin.
- Chemical degradation reactions involving pellicle proteins and carbohydrates, *e.g.* the Maillard reaction (which is what happens when fruit goes brown).

Intrinsic
- Enamel can be 'stained' from within by early childhood exposure to higher levels of fluoride (causing white or brown spots on the teeth).
- Some antibiotics, such as tetracycline, can stain developing permanent teeth (or the first set of teeth of an infant if taken by their mother when pregnant), appearing as yellow, brown or grey bands; minocycline use can sometimes lead to staining of permanent teeth in an adult.
- If a tooth is injured and exposed to internal blood, reddish brown iron may be left in the dentine, causing a darker yellow or brown tooth.

brushing and by various components added to oral healthcare products (discussed more fully in subsequent sections), although higher levels of staining may require cleaning by a dental professional. Cracks in the enamel may lead to deep-level staining that is more difficult to remove. Intrinsic tooth stain is not reversible with toothpaste intervention, although it may diminish over time or be managed with peroxide/bleaching.

4.2.8 Oral Malodour (Bad Breath)

Bad breath can be as simple as being due to something we have consumed (*e.g.* garlic, tobacco smoke), or due to the action of oral bacteria breaking down proteins in saliva to produce sulfur-containing chemicals such as mercaptans/thiols (responsible for the 'asparagus smell' in urine) and hydrogen sulfide (the 'rotten egg' smell of sewers). Oral malodour can be combated by regular oral hygiene to wash away the stagnant saliva and, if necessary, use of an antimicrobial product, most often found in a mouthwash. Bad breath is not necessarily always due to poor oral hygiene – there are a number of medical conditions that can cause it that should be discussed with a healthcare professional.

4.3 JUST WHAT ARE ALL THESE INGREDIENTS IN MY ORAL CARE PRODUCT?

4.3.1 Overview

Toothpaste formulation design has to take in several factors (see Box 4.3). Toothpastes have a basic function to remove food debris, pellicle stain and plaque, but additional functions, depending on need, can include caries prevention, breath freshening, teeth whitening, enamel building and combating dentinal hypersensitivity. As a product, a toothpaste has to have an appealing taste, feel and look and needs to be easy to use and store and have a long shelf-life. All of these factors need to comply with certain standards. The rules for selling oral care products comes under the EU Cosmetics Directive that defines a cosmetic product and requires toothpaste manufacturers to be able to substantiate any advertising claims.

BOX 4.3 IS EXPENSIVE TOOTHPASTE A SCAM?

Toothpastes used to just clean teeth, add fluoride to help with enamel structure and control dental caries and little else. Many people only need a simple toothpaste that provides anti-cavity protection with fluoride, but some oral challenges may require a bit more from a toothpaste formulation: help with specific oral conditions, such as gum disease or sensitivity; provision of a cosmetic benefit, such as whitening, and combinations of a number of functions.

Most toothpastes contain:

- fluoride (up to 1500 ppm) for caries control;
- abrasives (15–55%);
- humectants (5–70%);
- rheology modifiers (0.1–15%);
- surfactants (0.5–2.0%);
- flavours, sweeteners, colouring (0.1–2.0%);
- water (to make the formulation up to 100%).

Specialized toothpastes may contain ingredients for:

- stain removal (whitening);
- anti-sensitivity;
- enamel care;
- gingival health;
- calculus control;
- anti-plaque formation.

If a toothpaste makes a claim to do something ('caries prevention', 'anti-sensitivity'), then this claim must be substantiated *via* laboratory testing of the mechanism of action and clinical testing for efficacy and safety. These investigations are usually set up in the same way as they would be for a drug, so the process can be costly. Ingredients may also require specific sourcing along with needing specialized manufacturing components and standards. Add to this market testing for best flavour, look and feel (each new ingredient may change one or all of these aspects) and the expense of some toothpastes becomes clearer.

4.3.2 Fluoride

Fluoride is by far the best-known active ingredient of toothpaste, and has been added for many years (see Box 4.4). According to the FDI (Fédération Dentaire Internationale) World Dental Federation, 'regular use of fluoride toothpaste is scientifically recognized as a major means to reduce the prevalence and severity of dental caries and delay its onset'. Fluoride in the oral cavity can bind to calcium and phosphate during enamel remineralization to become hydroxyfluorapatite [$Ca_{10}(PO_4)_6(OH)F$], a compound that does not demineralize until the pH decreases to 4.5, compared with 5.5 for hydroxyapatite [$Ca_5(PO_4)_3OH$]. This pH is more difficult to attain and maintain with a typical daily diet but may still be impacted by frequent dietary challenges. As such, one of fluoride's main actions is as an 'anti-cavity' agent.

BOX 4.4 HOW LONG HAS FLUORIDE BEEN USED TO HELP WITH TOOTH DECAY?

- Mid-1700s: Note was made that ingestion of fluoride was reflected in higher levels in the teeth, making them 'stronger'.
- Early 1800s: It was observed that people living in areas where water was naturally fluoridated had lower levels of tooth decay, although in some areas this was also combined with high levels of unsightly fluorosis.
- Mid-1800s: Some dentists experimented with administering ingestible fluoride or fluorine to their patients.
- Late 1800s: The suggestion arose to add fluoride to water supplies, for development of both teeth and bone.
- 1934: The first patent for a fluoridated toothpaste, containing sodium fluoride (NaF), was filed by a US-based company; however, questions were raised around the safety of fluoride so it was not accepted at the time.
- 1930s–1950s: Research showed that fluorides were absorbed by tooth enamel and lowered its acid solubility.
- 1955/6: Stannous fluoride (SnF_2) was added to Crest® toothpaste, which also included the abrasive calcium pyrophosphate.

- 1960: Sufficient evidence was presented to the American Dental Association (ADA), *via* the use of clinical trials, that 'Crest® with Fluoristan' was accepted as 'an effective anti-caries dentifrice'.
- Early 1960s: Complaints came from customers that their SnF_2-containing toothpaste stained their teeth; investigation found this to be due to the action of tin (from the stannous part of SnF_2) on the pellicle.
- 1960s: Fluoride as sodium monofluorophosphate (SMFP) was proposed as another means of delivering fluoride within a toothpaste, without the staining properties of SnF_2. SMFP was first manufactured in 1924 and its caries-inhibiting effect was first examined in 1950. While initially incorporated into a toothpaste in the early 1960s, in 1969 Colgate® was the first SMFP formulation to be recognized for its anti-caries action by the ADA Council on Dental Therapeutics.
- 1975: The British Dental Association endorsed a number of fluoride-containing toothpastes to 'help prevent tooth decay'.
- 1994: The World Health Organization concluded that the use of fluoride in toothpaste was one of the main reasons for a decline in caries in countries where use was prevalent. Their International Scientific Panel declared that, when formulated in a silica-based toothpaste, NaF was more effective than SMFP for caries prevention.
- 2000s: SnF_2 formulations were optimized such that staining was limited by combining with stain-reducing materials; SnF_2 was also found to be useful as a dentine tubule blocking agent that could reduce/stop dentinal hypersensitivity.

Fluoride can be added to over-the-counter toothpaste up to a maximum level of 1500 parts per million (ppm) (EU Cosmetics Directive). It is recommended by the FDI World Dental Federation that teeth are brushed twice per day for a minimum of 2 min, especially last thing at night. Public Health England guidelines suggest that for the prevention of caries, a 'smear' of toothpaste 'containing no less than 1000 ppm fluoride' is used for children aged 0–6 years and a 'pea-sized' amount of toothpaste containing 1350–1500 ppm fluoride is appropriate for those aged 7 years and up, including adults.

> You're unlikely to see a shark with tooth decay as their teeth are naturally composed of fluorapatite [$Ca_5(PO_4)_3F$]

Fluoride comes in various forms, which have different properties that can depend on the other ingredients with which they are formulated. The most commonly used fluorides are sodium monofluorophosphate (Na_2PO_3F; SMFP), sodium fluoride (NaF), stannous fluoride (SnF_2) and amine fluoride (AmF). The choice of fluoride source revolves around formulation considerations that retain fluoride in an active state, *i.e.* it does not become bound to other ingredients and become inactive (see Section 4.3.3 for some examples of what can't be combined). In addition to considering other actions that the toothpaste is targeting (*e.g.* anti-sensitivity, anti-gingivitis), properties such as taste and overall cost of a formulation (*e.g.* in general, SMFP formulations are often the cheapest to make) need to be taken into account. A formulation may also include ingredients that work with fluoride's action; for instance, some early clinical studies suggested an increased efficacy in terms of limiting caries when 10% xylitol (an ingredient sometimes found in chewing gum and mouthwash) is added to an NaF–silica formulation. Xylitol is understood to work synergistically with fluoride to limit the amount of acid produced by plaque bacteria.

For those with a higher caries risk and people wearing orthodontic appliances, toothbrushing may be supplemented with a fluoride mouthwash (in Europe usually containing around 250 ppm fluoride), used a few hours afterwards to maximize continual oral exposure to fluoride. Alternatively, higher fluoride concentrations of 2800 and 5000 ppm are available in prescription toothpastes that can be prescribed by dental professionals for 'at-risk' patients.

Fluorosis – high deposition of fluoride within the teeth – occurs when the maturing permanent teeth of a child (0–6 years old) are exposed to higher levels of fluoride from multiple sources both external and internal, such as with fluoride tablets or drops, non-municipal water and toothpaste. Toothpaste exposure alone is not thought to lead to excessive fluorosis. It is important to note that with fluorosis, teeth are not damaged, but the enamel may have areas of opacities ('white spots') that

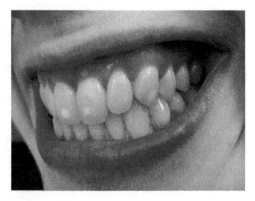

Figure 4.7 Mild fluorosis.
Reproduced from Josconklin, https://commons.wikimedia.org/wiki/File:Dental_fluorosis.jpg, under a CC BY-SA 3.0 license, https://creativecommons.org/licenses/by-sa/3.0/deed.en.

usually fade over time (Figure 4.7). Very rarely, the spots are darker than the enamel or the tooth is stained brown.

4.3.3 Abrasives

Abrasives have been used as tooth-cleaning elements since ancient Egyptian times when ingredients such as ground-up egg shells, pumice or ox hooves were added to a dental paste. Other ancient abrasives include snail shells and crushed bones, with more modern, eighteenth century, additions being brick dust, crushed china and ground cuttlebone.

An abrasive is usually an inert, insoluble material that can physically remove food debris, stained pellicle and plaque. These include hydrated silica (similar to sand), sodium bicarbonate (aka 'baking soda'), alumina (aluminium oxide), calcium carbonate (chalk, which also forms limestone from marine organism skeletons), calcium pyrophosphate and sodium metaphosphate. Some abrasives can also be used in toothpaste as thickening, anti-caking, bulking, opacifying, absorbance and dispersant agents.

The most commonly used abrasives are dental-grade silicas (SiO_2), also known as a component of sand, manufactured *via* gel, pyrogenic or precipitation processes. Silicas can constitute 5–15% of a toothpaste and the particle size can be varied to alter the abrasivity of a product. When a clear gel toothpaste is required, hydrated silica is used. Also commonly used is calcium carbonate ($CaCO_3$), which can comprise up to 40% of a toothpaste. Although useful, the calcium part can react with water-soluble fluoride, such as when it is formulated as NaF, to form insoluble calcium fluoride, rendering the fluoride inaccessible to the enamel; it is less reactive with fluoride as SMFP. Alumina [Al_2O_3] is a low-cost, good flavour compatibility ingredient that also works well with SMFP, although it can be highly abrasive and needs to be carefully controlled as it reacts with water-soluble fluoride to form an insoluble salt.

With all toothpastes, care must be taken to make sure that the abrasive component is not too abrasive, or else it may 'polish' teeth to the extent that damage is caused by removing layers of enamel. All toothpastes should comply with levels of abrasive action, which is something to enquire about when choosing a toothpaste. The abrasivity of a toothpaste can be measured according to its mechanical effects in a laboratory on dentine, known as 'relative dentine abrasivity' (RDA). Although there are some criticisms of this test (it is performed on natural tooth dentine, which can itself vary in hardness), an over-the-counter toothpaste should not go above an ISO standard RDA value of 250. The ability of an abrasive to remove the pellicle/stain can also be shown by its 'pellicle cleaning ratio', which is also based on a mechanical test.

4.3.4 Anti-sensitivity

There are two main ways to combat dentinal hypersensitivity: blocking the top of exposed dentine tubules or quashing the nerve activation associated with the occurrence of pain on stimulation. Such ingredients may be incorporated into a toothpaste or delivered separately in a mouthwash.

Along with its role as a way of delivering fluoride, SnF_2 has been shown since testing in the 1950s to also reduce dentinal hypersensitivity. The stannous (tin) salt works by physically blocking dentine tubules, a deposit that is then covered by the naturally forming dentinal smear layer and that is resistant to both acids and water. SnF_2 may be combined with sodium hexametaphosphate, which helps with the removal of dental stain, a known side effect of SnF_2 toothpastes, as it has a high affinity to pellicle proteins on enamel.

Strontium salts, which have been added to toothpastes to reduce sensitivity since the nineteenth century, are thought to work by complexing with dentine in and around the tubules, forming strontium apatite. They may also be combined with silica to form an acid-resistant silica–strontium deposit. The combination of arginine and calcium carbonate forms a dentine-like calcium and phosphate layer in and on top of dentine tubules.

The bioglass calcium sodium phosphosilicate (CSPS), used since the 1970s as a bone-building agent (see also Section 4.3.8), provides the natural building blocks of tooth enamel from a toothpaste. It can form a deposit of some of its constituents – calcium, phosphate and silica – over and into dentine tubules, forming a hydroxyapatite-like layer (Figure 4.8). This is both acid and water insoluble and resistant to mechanical removal by a toothbrush or abrasive. The deposit may also contain titanium, if this is formulated into the toothpaste.

Oxalates, which are formed of calcium and oxygen, can react with calcium ions in saliva to form insoluble calcium oxalate crystals. Although, like the above, they can be deposited into dentine tubules, they are acid soluble and hence can be easily washed away in lower pH environments. They are usually used in the form of potassium or ferric salts.

Nerve desensitizing agents are centred around potassium salts such as nitrate, chloride or citrate. These are thought to 'depolarize' the nerve fibres at the base of dentine tubules such that

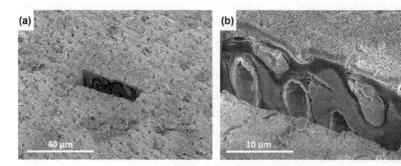

Figure 4.8 (a) Low- and (b) high-magnification scanning electron microscope images of a cross-section showing a calcium sodium phosphosilicate (CSPS) layer across the dentine surface, occluding dentine tubules.
Reproduced from ref. 4 with permission.

any signal is not transduced. Pain signalling partly relies on a gradient of potassium with high concentrations inside the nerve fibre and low concentrations outside. The flow of potassium ions from internal high to external low pools (and the reverse for sodium) is what moves the pain signal along. If there is a high level of potassium outside, equal to that inside, the 'pain signal' cannot be generated as there is no flow between inside and outside. In general, desensitizing agents take a longer time to work than those that can immediately physically block dentine tubules (*i.e.* after a number of days' use) as the effect on nerve endings may be gradual while potassium levels build up to be equal to that inside the nerve fibre.

For those where dentinal hypersensitivity does not respond to over-the-counter treatments and good oral hygiene practices, a dental professional may use a treatment such as a fluoride varnish, also containing calcium and silica, that can remineralize enamel and dentine and form a short-term protective layer over dentine tubules.

Like all toothpaste ingredients, those that can aid sensitivity need to be tested in clinical trials. However, there are nearly always problems with conducting trials involving perception of pain.

Pain as a single entity is a very hard thing to measure. It can change throughout the day in both strength and form, *e.g.* a tooth that is sensitive on waking may be less so later in the day and may be more sensitive to cold air at lunchtime and a toothbrush's touch at night. Pain is also connected with the emotional feelings it elicits, which can alter the level and experience of pain over the course of a clinical trial.

As such, similarly to all pain studies, anti-sensitivity studies are always carried out with the understanding that participants in whom pain is reduced over the course of a trial may be experiencing a placebo effect, where just being told they are receiving a pain-reducing treatment leads to a reduction in pain, or a Hawthorne effect, where they alter their behaviour (knowingly or not) due to being observed in a trial.

Efforts to counteract these problems include carrying out laboratory testing of anti-sensitivity materials to show their physical effects and using two different types of pain stimuli in a clinical trial, adding participant rating scales to those of an independent observer and requiring a threshold to be passed in terms of pain reduction, *e.g.* a 20% change from the initial (baseline) measurement before the trial began.

4.3.5 Stain Removal/Whitening

The promise of tooth whitening has become big business in recent years. Achieving it can be roughly divided into three ways: (i) *via* over-the-counter products for everyday stain removal, such as *via* a toothpaste and/or mouthwash, (ii) *via* over-the-counter occasional use products, such as bleaching/whitening strips or dental trays, and (iii) *via* the use of professionally applied bleaching products, carried out in a clinic (Figure 4.9).

Whitening toothpastes may achieve that 'label' by the addition of both stain-targeting abrasives and chelating agents. Abrasives may be harder and be present at a higher concentration than in regular family toothpastes (closer to the limit of RDA = 250). Care must be taken when a higher abrasivity toothpaste is used by someone who has thinning enamel, receding gums and exposed dentine, as such toothpastes may increase tooth and tissue loss. Thus, while a higher abrasivity toothpaste may remove more stain, it may also remove the top layer of enamel and, at the gum line, dentine, at the same time. Although the abrasive properties of toothpastes can be adjusted through careful formulation approaches to target a measured abrasivity without losing its

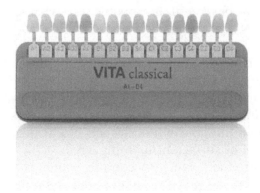

Figure 4.9 VITA classical A1–D4® shade guide, which may be used by a dental healthcare professional during in-clinic whitening. Image credit © VITA Zahnfabrik.

freshening/cleaning function, they may be less effective at removing dental stain. Properties often adjusted include size (a smaller abrasive has a larger surface area), shape (round or irregular), composition (hard, medium or soft compound) and concentration, similarly to the different types of coarse to fine sandpaper.

- Hard/high stain removal ingredients: perlite, alumina
- Medium hard/medium stain removal ingredients: hydroxyapatite, calcium pyrophosphate, hydrated silica
- Soft/low stain removal ingredients: sodium bicarbonate, calcium carbonate, brushite (dicalcium phosphate dihydrate)

One abrasive that has been used for many decades is sodium bicarbonate ($NaHCO_3$; aka baking soda). It is something of a 'love it or hate it' ingredient owing to its mildly salty taste and slightly gritty feel. Although it is considered a 'soft/low stain removal' abrasive, during use sodium bicarbonate can be distributed to the harder to reach areas of the mouth that may be missed when brushing, allowing the gritty element to work throughout the oral cavity.

Blue covarine works in an entirely different way. Its blue colour is used to counteract the yellow of tooth stain so that teeth appear whiter without removing pellicle or altering enamel. This ingredient can be useful for people for whom other tooth-whitening/bleaching products are not suitable. As the pellicle is partly made of protein, it can also be broken down by enzymes/proteases such as papain (found in papaya) or bromelain (found in pineapples). Surfactants also have some light stain-removing properties as they can get rid of hydrophobic ('water-hating') compounds in the same way that stains can be washed away by soap. Nearly all toothpaste/mouthwash products generally contain some level of surfactant/foaming agent.

Oral care products can also contain agents that limit stain deposition, such as polyphosphates and sodium citrate, that help prevent calculus formation and can bind chromogen ions from food, beverages, *etc.*, preventing their deposition. Polyphosphates include sodium/potassium pyrophosphate, sodium trimetaphosphate and sodium hexametaphosphate. Calcium phosphates can adhere to enamel and prevent plaque formation and bacterial attachment.

Finally, charcoal has recently been marketed as a stain-removal and tooth-whitening ingredient in toothpastes and powders. It is reported to be an ancient and widely used ingredient. The fine powder of modern-day oxidized/activated charcoal can be made from the controlled reheating or chemical treatment of nutshells, olive pits, coconut husks, peat, wood and bamboo. Currently there is insufficient scientific evidence to back up significant claims, be they cosmetic (stain removal/whitening) or therapeutic (antimicrobial action, caries reduction). This means that, to date, insufficient adequate clinical trials have been performed and published that have shown products containing charcoal to be significantly better than products without it. As noted previously, claims in advertisements should be substantiated as a requirement for toothpastes marketed in the EU. Lastly, because charcoal is a relatively 'new' or 'on-trend' ingredient in commercial toothpastes, less information is known about it. Some academics have raised concerns over regular twice-daily oral charcoal use as it might have a 'possible health risk' if the preparation of the charcoal includes polycyclic aromatic hydrocarbons, which are considered to be carcinogenic (cancer-inducing). Worries also centre around the ability of highly absorptive activated charcoal to sequester active toothpaste ingredients including fluoride ions, thus rendering them useless. Finally, concerns have been raised over the abrasivity of some charcoal preparations where the whitening performance is gained through high abrasivity (RDA up to 250).

Peroxide is only allowed at concentrations up to 0.1% in EU cosmetic toothpastes. When used at higher concentrations in professionally applied products, it is the most commonly used bleaching agent for the teeth. It comes in the form of hydrogen peroxide (H_2O_2), calcium peroxide (CaO_2), sodium percarbonate ($Na_2H_3CO_6$) and magnesium peroxide (MgO_2). Activated chlorine dioxide as a source of hydrogen peroxide has also been proposed, although it is not widely used. These systems work by oxidizing organic chromogens, a chemical reaction in which an electron is lost from an atom, altering the properties of the atom.

The result is that organic chromogens that usually absorb light in the visible region of the spectrum become oxidized such that light is no longer absorbed and they become non-coloured and 'invisible' – hence why this treatment is referred to as 'bleaching' and is a different approach to an abrasive stain-removal agent.

Dental professionals may also offer whitening treatments that are a combination of H_2O_2 and light. This is thought to increase the rate, but not the amount, of chromogen oxidation by increasing the energy/temperature. Such treatments can produce a tooth colour that is lighter than the natural colour as the peroxide penetrates into the top layer of enamel. One noted side effect of all bleaching approaches, although usually transient, can be dentinal hypersensitivity.

4.3.6 Anti-gingivitis

As already discussed, control of gingivitis is essential to help prevent the transition to destructive, irreversible periodontitis. Anti-gingivitis agents involve those that can help limit gum inflammation and those that reduce levels of plaque or plaque build-up.

One of the most investigated agents for plaque control is chlorhexidine in a mouthrinse. This is a broad-spectrum antibacterial agent that is retained for several hours in the tissues of the oral cavity, leading to a prolonged effect. However, chlorhexidine is a problematic ingredient as it reacts with many other ingredients in a toothpaste formulation to inactivate its beneficial properties. When formulated correctly, known side effects of its mode of action are that it can stain teeth brown with prolonged use, alter taste perception and increase calculus deposition. Hence it is mostly only used in short-course mouthwashes as a therapeutic product (often classified as a medicine), when the need outweighs the side effects, and with advice on how to combat adverse events including limited use and adjunctive use of a whitening toothpaste.

Triclosan has been shown to have anti-inflammatory properties on tissue in addition to being an antimicrobial agent. It is often formulated with zinc citrate or an organic copolymer called Gantrez™, which helps it remain in the mouth for longer. Its use has been questioned, however, as it has been shown to accumulate in the aquatic environment, where it may end up after going down the sink. There are also concerns that its use can lead to antimicrobial resistance and all major toothpaste manufactures have ceased using it.

SnF_2, along with its anti-cavity and anti-sensitivity properties (see the relevant sections), also has a role as an anti-gingivitis agent. Tin can also be formulated as stannous chloride ($SnCl_2$) for this use. It is thought to act as an antimicrobial agent, thus lowering inflammation associated with bacterial growth, in addition to having plaque-lowering/growth inhibition properties.

Another multifunctional toothpaste ingredient is sodium bicarbonate. Although used for centuries as a mouth-cleansing compound, it is only recently that clinical trials have confirmed its use as an anti-gingivitis agent. These showed that it can reduce indices of gingivitis such as bleeding on provocation. Laboratory studies have found that sodium bicarbonate can both physically and chemically remove mature dental plaque, which is reflected in clinical studies where plaque occurrence was reduced in both toothbrush-accessible and more inaccessible sites.

Finally, a recent 'on-trend' idea is the use of 'oil pulling' therapies. Like many trends, it actually has roots back to traditional medicine, most notably in India as an Ayurvedic treatment. The basic principle is that swishing with an oil may be effective in suspending oral debris and bacteria for easy removal from the mouth through expectoration (spitting). Oil pulling may work simply due to the mechanical action on disrupting less mature plaque that swilling entails. One of the most commonly used oils – coconut oil – may also work to reduce bacterial load as it includes lauric acid, which has been shown to have some antibacterial properties. Recent clinical studies of coconut oil pulling have shown that it may indeed lead to some reduction in plaque and gingivitis, with scores equivalent to those found following the use of chlorhexidine. Studies are also looking into the efficacy of sesame and sunflower oils. However, the basic question of how to provide effective caries control in a highly processed sugar/carbohydrate remains unresolved, so it is likely that oil pulling should be additional to toothbrushing with a fluoride toothpaste.

4.3.7 Calculus Control

As discussed in Section 4.2.2, calculus is formed mostly of calcium and hydroxyapatite, meaning that its composition is similar to that of enamel. This can be problematic when developing an anti-calculus toothpaste as it means that it is not as simple as adding a demineralizing agent. Calculus is best first tackled by a thorough cleaning ('prophylactic' treatment) by a dental professional that removes any calculus on the teeth. A toothpaste with a crystal growth-inhibiting ingredient that helps limit calculus build-up without harming enamel can then be used. Such ingredients include pyrophosphates (*e.g.* tetrasodium pyrophosphate, tetrapotassium pyrophosphate or disodium dihydrogen pyrophosphate), diphosphonates and zinc citrate. These react with calcium and phosphate before they are deposited on the teeth or by removing plaque and preventing its build-up.

4.3.8 Enamel Care

As discussed, at lower levels of enamel demineralization, fluoride can help with enamel remineralization; however, the amount of enamel deposited following use is low. More recently, 'bioglasses' (aka bio-active glasses), most notably in the form of CSPS (calcium sodium phosphosilicate), have been added to toothpastes to help provide the natural building blocks of enamel. However, there are limited data available on any additive benefits over and above those provided by fluoride alone and CSPS is primarily an anti-sensitivity agent due to its dentine tubule occlusion properties.

4.3.9 Other Ingredients

Humectants help stop toothpaste from drying out by binding with water molecules hence reducing the loss of water from the surface. They are also added to improve the taste of a toothpaste and act as preservatives by limiting the availability of water to microorganisms. Glycerin and sorbitol are the most often used humectants; they reduce the availability of water, improve taste and texture and have some preservative functions. Other humectants include sugar alcohols (*e.g.* xylitol) and glycols [*e.g.* propylene glycol, poly(ethylene glycol)].

Surfactants (detergents or surface-active agents) work by lowering the surface tension between a toothpaste and the pellicle and tooth (for more information on surfactants, see Chapter 2). They are responsible for the foam of a toothpaste, which helps to distribute the ingredients throughout the oral cavity, and can add to the cleaning properties of a product. The most commonly used surfactant is sodium lauryl sulfate (SLS), synthesized from naturally occurring alcohols. Alternative surfactants include sodium methyl cocoyl taurate and cocamidopropyl betaine. The levels and types of surfactants that can be used in toothpastes are limited as they can be oral irritants and taste bitter (see Table 4.1).

Rheology refers to how a substance flows. In a toothpaste, this factor is controlled by a number of gelling and thickening agents that alter structure, viscosity and stability. The most common, which are gelling agents, are cellulose/hydroxyethylcellulose gum (from plant cell walls), xanthan gum (from fermented sugars) and carrageenan (from seaweed). Other agents include carbomers, bentonite, hectorite and silica/hydrated silica (used as a thickener).

Flavours, sweeteners and colouring help distinguish a toothpaste according to consumer preference. They must all be 'food approved' for use. Flavourings include peppermint, spearmint and clove oils, eucalyptol, cinnamon, aniseed, wintergreen, menthol, methyl salicylate and anethole, along with synthetic

Table 4.1 Inactive ingredients that may be included in a toothpaste formulation.

Type	Ingredients
Abrasives	Alumina, calcium carbonate, dicalcium phosphate dihydrate, hydrated silica, silica
Humectants	Glycerin, sorbitol, xylitol
Surfactants	Cocamidopropyl betaine, sodium lauryl sulfate, sodium methyl cocoyl taurate
Rheology modifiers	Bentonite, carbomer, carrageenan, cellulose gum, hectorite, hydrated silica, hydroxyethylcellulose gum, silica, xanthan
Flavourings	Anethole, carvone, eucalyptol, *Eugenia caryophyllus* (clove) flower oil, *Mentha piperita* (peppermint) oil, *Mentha spicata* (spearmint) oil, menthol, methyl salicylate
Sweeteners	Sodium saccharin, *Stevia rebaudiana* extract, sucralose, xylitol
Colourants	Beetroot extract, Blue No. 1, carotenoids, chlorophyll, Red No. 40, titanium dioxide

fruit flavours. Sweeteners, as would be expected in a toothpaste, cannot be those used by oral microorganisms as food. They include sodium saccharin, xylitol, stevia and sucralose (a non-fermentable sugar), added not only to enhance the flavour itself, but also to combat the bitterness of other ingredients such as some surfactants. Colourings include food dyes such as Blue No. 1 (a synthetic organic compound); Red No. 40 (from sodium, calcium or potassium salts) and plant-based extracts such as chlorophyll green, beetroot pink and carotenoid orange. More recently, 'sparkles' in the form of mica (a mineral similar to silica) have been added, especially to children's toothpaste.

4.3.10 Mouthwash

Many of the ingredients that make up a toothpaste can also be present in a mouthwash; however, the primary function here is to freshen the mouth and breath. In general, mouthwashes contain fewer ingredients than toothpastes. Many still include ethyl alcohol (5–25%), functioning as both a preservative and a flavour-enhancing agent; a humectant (sorbitol or glycerin), to aid 'mouth feel', increase viscosity and, to a lesser extent, add sweetness; a surfactant, for stability and some foaming; and flavouring, colouring and sweetening ingredients. Active ingredients can include fluoride, hydrogen peroxide, potassium, pyrophosphate and chlorhexidine ions to aid with, respectively, caries protection, whitening, anti-sensitivity, anti-calculus and anti-microbial effects.

4.3.11 How Are Claims for Oral Care Product Performance Substantiated?

Although the basic formulation of a toothpaste has not changed for decades, tweaks are made to aid the purpose of a particular blend. This involves testing of new formulations both in the laboratory and in clinical trials, for efficacy and safety.

- Remineralizing ingredients
    - In the laboratory, discs coated with hydroxyapatite, squares cut from extracted human teeth (usually 'wisdom teeth') or animal teeth (usually cows) are swirled in

artificial or natural saliva, then pH-lowering agents (such as grapefruit juice) are used to cause demineralization of the hydroxyapatite/enamel.
- Demineralization can be viewed and recorded under a microscope and measured using very accurate depth analysis.
- The individual ingredient, or toothpaste formulation, proposed to aid remineralization is added over the hydroxyapatite disc/enamel sample in a new bath of saliva and any rebuilding of enamel is measured as before.
  - Enamel squares can also be mounted onto mouthpieces and worn by an individual for a specified amount of time so that the action of a toothpaste formulation can be tested in the more real-world setting of a person's mouth.
- Anti-sensitivity
  - Squares of dentine from extracted teeth can be exposed in the laboratory to the ingredient/toothpaste formulation proposed to block dentinal tubules.
    - The degree of dentine tubule blocking can be seen *via* a microscope or can be tested by analysing fluid flow through the tubules.
    - Dentine squares can also be mounted onto mouthpieces as tested for tubule occlusion '*in situ*' in a person's mouth.
  - In clinical trials, the degree of dentinal hypersensitivity can be tested by stimulating the tooth before and after treatment.
    - Sensitivity can be assessed by touch *via* a probe that applies a defined pressure; the pressure at which pain is elicited is then noted.
    - Sensitivity can also be assessed by the use of cold air *via* a dental air syringe, with the reaction to this stimulus rated by a suitably trained dental professional.
    - The participant can rate their pain level on a simple scale or by using a specialist scale describing particular aspects of the experience of dentinal hypersensitivity.
    - The toothpaste formulation will be used for a defined period (from single use to many weeks), during which the participant is regularly retested.

- Whitening/stain removal
  - Discs of hydroxyapatite prestained with, for example, black tea or squares of naturally stained/prestained enamel can be exposed to the stain-removal/whitening component in the laboratory.
    - If the proposed mechanism is mechanical, such as a new type of toothbrush or bristle, the discs will be mounted on a multi-head scrubbing machine.
    - If the mechanism is physical or chemical, such as with abrasives or peroxide, the scrubbing machine may also be used (comparing with/without the toothpaste) or a slurry of the toothpaste may be swirled with the discs/squares.
  - Clinical studies can involve people with highly stained teeth, either naturally so or by using a 'forced stain' model where participants swill with chlorhexidine then tea.
    - The action of a product may be noted by how well it whitens/removes stain or else teeth may first be professionally cleaned to remove all stain and the action to limit stain build-up is noted.
  - Stain can be measured in a number of ways.
    - Matching to a standard set of 'teeth' known as the VITA shade guide (Figure 4.9).
    - Using a 'stain index' to rate stain degree (how much) and extent (how far).
    - Digital detection of the exact shade can be used in laboratory studies using discs or extracted enamel and in the clinic directly on teeth or by using pictures taken under controlled lighting conditions.
- Gingival inflammation
  - Measures of the degree of inflammation use rating scales that rank colour (pale pink to red), swelling and extent of such.
  - Poking the gums at a number of defined spots with a sharp probe allows the investigator to see how many 'bleeding spots' there are in a person's mouth, how long it takes for a spot to bleed and how long the bleeding lasts.
  - As gingival inflammation can be caused by the presence of plaque bacteria, samples may also be taken to ascertain what bacteria are present and at what level; laboratory

testing may also be used to analyse directly the action of an ingredient/formulation on bacterial growth.

4.4 CONCLUSION

When looking at the list of ingredients in a toothpaste and mouthrinse, it may seem quite long for a seemingly simple product. As we have seen though, not only must a toothpaste function to help remove stain and prevent decay, additional benefits can include ingredients to combat plaque build-up, control dentinal hypersensitivity, limit gingivitis and aid in teeth whitening and calculus control. These ingredients often have chemical and physical properties that mean the formulation of them together in a toothpaste is complicated, especially when it also has to be something that works well and does not harm the oral environment of teeth, gums and soft tissue, along with having a consumer-acceptable taste and feel.

Finally, how important is your oral health beyond your mouth? Recent research has demonstrated that the worse your oral health, the more likely you are to develop a host of potentially life-threatening conditions that can impact, sometimes severely, your quality of life (*e.g.* dementia, heart disease, diabetes, obesity). Hence, brushing your teeth twice per day, every day, for 2 min each time with a fluoride toothpaste that you like has never been more important for your overall health.

FURTHER READING

I. L. C. Chapple, F. Van der Weijden, C. Doefer, D. Herrera, *et al.*, Primary prevention of periodontitis: managing gingivitis. *J. Clin. Periodontol.*, 2015, **42**(Suppl. 16), S71–S76.

M. Epple, F. Meyer and J. Enax, A critical review of modern concepts for teeth whitening. *Dentistry J.*, 2019, **79**, 1–13.

J. D. Harvey, Periodontal microbiology. *Dental Clin. North America*, 2017, **61**, 253–269.

S. Mason, Dental hygiene, in *Poucher's Perfumes, Cosmetics and Soaps*, ed. H. Butler, Kluwer Academic Publishers, UK, 10th edn, 2000, ch. 7, pp. 217–253

J. M. ten Cate, Contemporary perspective on the use of fluoride products in caries prevention. *British Dental J.*, 2013, **214**, 161–167.

C. van Loveren, R. M. Duckworth, Anti-calculus and whitening toothpastes, in *Toothpastes, Monographs in Oral Science*, ed. C. van Loveren, Karger, Basel, 2013, vol. 23, pp. 61–74.

G. L. Vogel, Oral fluoride reservoirs and the prevention of dental caries, in *Fluoride and the Oral Environment, Monographs in Oral Science*, ed. M. A. R. Buzalef, Karger, Basel, 2011, vol 22, pp. 146–157.

N. West, J. Seong and M. Davies, Dentine hypersensitivity, in *Erosive Tooth Wear, Monographs in Oral Science*, ed. A. Lussi and C. Ganss, Karger, Basel, 2014, vol. 25, pp. 108–122.

REFERENCES

1. D. F. Kinane, P. G. Stathopoulou and P. N. Papapanou, *Nat. Rev. Dis. Primers*, 2017, **3**, 17038.
2. R. Levine and C. Stillman-Lowe, in The Scientific Basis of Oral Health Education, BDJ Clinician's Guides, *Dental Erosion and Erosive Tooth Wear*, Springer, Cham, 2019, pp. 49–56.
3. C. Ganss and A. Lussi, *Monogr. Oral Sci.*, 2006, **20**, 32–43.
4. J. S. Earl, R. K. Leary, K. H. Muller, R. M. Langford and D. C. Greenspan, *J. Clin. Dent.*, 2011, **22**, 62–67.

CHAPTER 5

You Against the World! – The Science Behind Skin and Skincare Products

ROBIN PARKER,* NICHOLA ROBERTS* AND MONIQUE BURKE

Acheson & Acheson, Meridian Business Park, Trowbridge BA14 0BP, Wiltshire, UK
*Email: robin.parker@acheson.co.uk; nichola.roberts@acheson.co.uk

5.1 THE SKIN – WHAT EXACTLY DOES OUR SKIN DO?

When people are asked to name five bodily organs, skin is rarely on the list. Why would this be when it's actually the largest organ and the one most visible? For some reason, it's often not even thought about as an organ at all. This may be because its function is seen to be less exciting and dynamic than that of many of the other 'show-off' organs. After all, it's just a cover, isn't it? A simple protective layer designed to stop the more interesting, exciting stuff falling out?

Thankfully for cosmetic scientists, dermatologists and an increasing number of other scientific disciplines, this isn't true.

Discovering Cosmetic Science
Edited by Stephen Barton, Allan Eastham, Amanda Isom, Denise McLaverty and Yi Ling Soong
© The Royal Society of Chemistry 2021
Published by the Royal Society of Chemistry, www.rsc.org

Skin is much more complicated than first meets the eye and deserves a far higher billing on the all-time favourite organs list. Skin is, quite simply, remarkable!

We could argue that no other organ has the power of our skin, it literally helps to define us. When any human lays eyes on another human, the first thing they will see is their skin – the colour, the evenness of tone, the smoothness, whether it is shiny or dull. Within seconds we can tell the ethnicity and the approximate age of the individual just by a glance at their skin. Subconsciously we can make a judgement about general health from the appearance of skin, subtle colour variations often being associated with underlying health issues. Conversely, when we feel that our skin doesn't look great it can have a huge impact on our self-esteem and confidence, and this in turn can potentially lead to other health problems. Whether we realize it or not, we can even tell how the other person feels about the sight of us by colour changes to skin as we communicate, or possibly by visible moisture on the skin surface. When human interactions move beyond just sight and touch is involved, the power of skin takes on a whole new level. The sensory nerves contained in the skin give and receive messages that lead to closer bonding (or possibly not). It is now well understood that these powerful sensory messages are among the first signals we receive after birth and play a vital role in the bonding of mother and child.

It is true that the primary function of the skin is protection – keeping infection out, keeping foreign bodies out, protecting against physical damage, helping to regulate body temperature. But it is also so much more than this and in turn it deserves our protection as we owe so much to it. Skincare products help to keep our skin remarkable.

5.1.1 Skin Deep – What Is Beneath the Surface?

With a surface area of between 1.5 and 2 m^2 and weighing in at around 15% of our body weight, the skin is one of the largest bodily organs. Only the liver challenges it for the number one spot. It was once thought to be an impermeable barrier that had evolved to protect the rest of the body, keeping everything out and allowing nothing in. We now know that skin is amazingly complex, made of four distinct layers, with some of these layers subdivided into further layers, all with specific functions and roles (Figure 5.1).

The first thing to understand about the skin is that it is dynamic. The surface, the visible part, is known as the stratum corneum (SC) and itself consists of several layers of skin cells (corneocytes) that have migrated up through the skin and living a short life from birth to death of around 30 days depending on the age of the person. The corneocytes are held together by complex proteins called corneodesmosomes, which act like microscopic rivets. As will be described later, the corneocytes are surrounded by specialized lipids – fatty materials that help form the skin barrier. After the 30 day transit period, the corneocytes start to dry out, they flatten, the nuclei disappear and the corneodesmosomes begin to break down. Eventually the cells are shed, a process known as desquamation, and join the rest of the dust in our surroundings (Figure 5.2).

New skin cells (keratinocytes) begin life in the base of the epidermis and, as they develop and grow, they move up through the epidermis, pushed up by new cells continuously produced below. At the base of the epidermis lie special cells known as melanocytes. These cells produce melanin, the brown pigment that defines our skin colour. The melanocytes have finger-like projections known as dendrites that deliver the melanin into the keratinocytes as they develop and travel up to the stratum corneum.

Underneath the epidermis is the dermis, a complicated layer that acts as the control centre for the skin and contains many of the main structural elements that give our skin strength and elasticity. Cells called fibroblasts are found in this area and these control the production of the extracellular matrix, comprising the structural tissues collagen and elastin along with water-binding glycoproteins such as hyaluronic acid.

Sitting within the dermis, in the hair follicles, are sebaceous glands, which produce sebum – the oily, waxy material that coats the skin. The greatest concentrations (reportedly as high as 400–900 glands cm^{-2}) are found on the scalp, face, upper chest and shoulders – these are the areas where greasy skin appears when there is overproduction of sebum. As we will learn later, spot outbreaks and acne can result, but the positive benefit of sebum is that it helps repel water and, with small amounts of sweat, forms a lightly acidic coating across the skin.

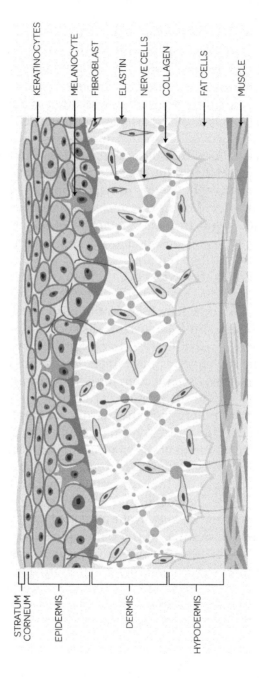

Figure 5.1 Cross-sectional representation of skin.

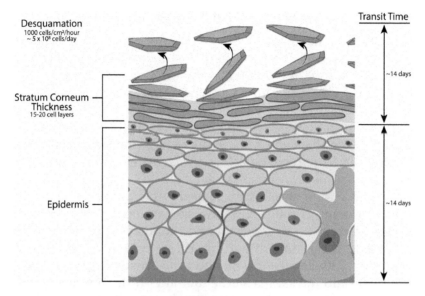

Figure 5.2 Desquamation of skin cells.

The deepest layer, the hypodermis, consists of connective tissue, fat and blood vessels.

Together, these complex layers all work to give our skin the remarkable properties that we so rely on – a protective barrier that is tough yet flexible, a cushion against minor knocks, an organ delicate enough to allow us to feel the touch of a feather (see Box 5.1).

BOX 5.1 THICKNESS OF THE SKIN ON OUR BODY

The skin around the eyes is particularly delicate and in fact the eyelids themselves are the thinnest skin on the body at around 0.05–0.1 mm; the thickest skin is on the feet, where it is around 1.5–4.0 mm.

The palms, particularly in certain areas, have much thicker skin than most other parts of our bodies; this gives it the strength and resilience required for constant holding and gripping. However, on the backs of our hands the skin is relatively thin with very little hypodermis.

5.1.2 We Are Not Alone – What's On The Surface?

The stratum corneum, the outermost layer of human skin, is home to many species of bacteria, fungi and viruses all living in harmony on our skin surface and within the stratum corneum (Figure 5.3; see Box 5.2). The two most commonly used terms to describe this, 'skin microbiome' and 'skin microbiota', are often used interchangeably but do in fact have subtle differences in meaning. We won't go into these subtleties here and will use the term 'skin microbiota'.

Traditional microbiological analysis using swabs and agar plates is not sensitive enough to detect the diversity of species on skin but it is now possible to map the microbiota on the skin surface by using genetic analysis of skin swabs. The hundreds of species present can be detected and the relative concentrations calculated. At the time of writing, these techniques are still in their infancy and are expensive and time consuming to perform.

In the womb the skin is sterile, but after birth and during the birthing process environmental microbes rapidly start to colonize the stratum corneum, eventually developing into a complex microbial ecosystem.

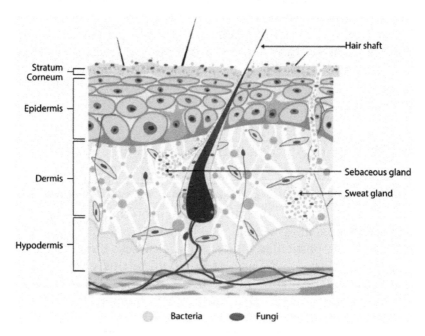

Figure 5.3 Distribution of bacteria and fungi on and within skin.

> **BOX 5.2 MICROORGANISMS ON OUR SKIN**
>
> Hundreds of species of bacteria can be found on the skin, but the most common and best known are the following:
>
> *Staphylococcus epidermidis*
> *Staphylococcus aureus*
> *Cutibacterium acnes* (formerly *Propionibacterium acnes*)
> *Pseudomonas aeruginosa*
> *Corynebacterium* spp.
>
> The human microbiota includes bacteria, fungi, mites and viruses and each square centimetre of the skin is home to approximately one million microorganisms!

Most species are harmless or even beneficial, playing a role in modulating the immune system in the skin. However, an unbalanced microbiota is linked to many skin conditions such as eczema, dandruff, acne and dermatitis and, occasionally, can result in serious infections.

Many factors can affect the microbiota, including pH changes, sebum content, barrier function, hydration, lifestyle, climate, gender, age and ethnicity. Also, different areas of the skin have very different populations of bacteria and fungi. *Cutibacterium* species, for example, are more plentiful where sebaceous glands are present and dry areas of the skin have the greatest diversity of species while having the lowest total number of bacteria. We will come back to *Cutibacterium* later in the chapter.

In the last few years, it has become accepted that when in healthy balance, commensal species of microorganisms are an important part of the natural immune system of the skin.

By competing for space and food sources, no one species can become dominant and, furthermore, some bacteria directly restrict the growth of competitors *via* production of antimicrobial peptides (AMPs) that can inhibit the reproduction of other, potentially harmful, bacteria.

Helping to keep this healthy balance in check is one of the most important functions that skincare can deliver and it is an exciting new area of research for cosmetic science. One of the most unusual ways to include this function in skincare is to formulate with a prebiotic ingredient. These ingredients act as specific food sources for the bacteria and have been shown to have moderate effects on the population balance.

Addition of probiotic ingredients – live bacteria – is rare. Clearly this is not practical for a finished formulation that needs to be safe and stable for several years. An alternative way of using probiotic technology is to formulate with active ingredients that have been synthesized using live bacteria rather than using the bacteria themselves. These have in some cases been shown to offer specific skincare benefits.

The dilemma for all formulators targeting the microbiota is that aqueous-based skincare formulations must be preserved to prevent spoilage (see Chapter 9). The necessary microbiological testing demonstrates that the formulations have a powerful effect against microorganisms accidentally introduced into the formulation during use – particularly important for pathogenic organisms hanging around in your bathroom! So, the challenge is how to preserve the skincare product successfully but at the same time have a minimal effect on the natural microbiota once applied to the skin. Demonstrating that this has been achieved is currently not easy, but the techniques for analysing the microbiota described above may start to become more accessible and less expensive.

5.2 ONE SIZE FITS ALL? – ALL SKIN IS DIFFERENT

So far, we've assumed that all skin is the same. The fact that all skin is different is another thing that makes skin so amazing. Variations between individuals, between different body sites in the same individual, seasonal changes in any individual and long-term changes as they age – these and many other factors are what keep cosmetic scientists on their toes when designing products. Here are just a few with which you may be familiar.

5.2.1 How and Why Does the Skin's Appearance Change with Age?

A basic understanding of the ageing process of skin is important for any cosmetic scientist. We know that ageing cannot be reversed or even slowed, but well-designed cosmetic skincare products can influence the appearance that these changes produce. To keep skin looking and feeling younger and healthier at any age is the objective.

Changes start to occur as early as our 20s. The skin cell turnover starts to slow gradually. In our 30s, the natural desquamation process also gradually slows, the cell turnover continues to decline and in the dermis, the rate of collagen production begins to slow. The structural changes that result begin to become visible. In our 40s, the epidermis begins to thin and collagen production continues to decline. Collagen and elastin can become further changed and degraded through a process called glycation. Lipid production slows, leading to less protection at the stratum corneum, and the glycoproteins in the dermis including hyaluronic acid begin to decline. By our 50s, all these processes continue apace along with a reduction in the fat layer beneath the skin and a less controlled production of melanin, leading to age spots and uneven pigmentation. These cumulative changes all lead to very visible effects by this stage. The loss of structure causes sagging and wrinkles become more defined. In female skin, the hormonal disruption caused by menopausal changes that occur in the 50s also has a significant impact. Figure 5.4 illustrates these changes as we move from our 20s to our 50s and beyond.

All these changes to skin occur naturally as we age and are largely determined by genetics. Lifestyle and general health factors can influence when these changes occur and how dramatic they are, but ultimately they are predetermined. This type of ageing is known as *intrinsic* skin ageing.

Alongside this natural ageing process, a number of external factors can and do produce additional ageing effects. These are known as *extrinsic* ageing factors and they are the ones that individuals can have the most influence on through changes to lifestyle and the adoption of good skincare routines.

UV radiation has long been shown to have a dramatic influence on the skin ageing process, known as photoageing. UV radiation is part of the electromagnetic spectrum and different wavelengths of light have different effects on skin. Broadly, UVA radiation penetrates deeper into the skin and can cause longer term damage; UVB radiation, on the other hand, penetrates less

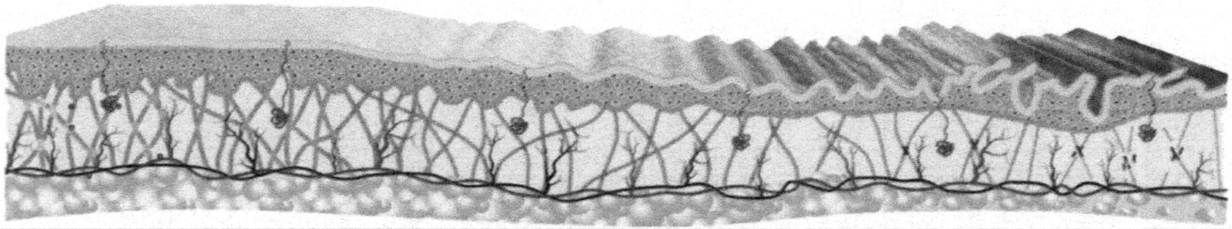

20's	30's	40's	50+
Cell turnover in the epidermis begins to slow down. Hormone imbalance from teen years settles. Skin can be at its optimum adult appearance.	Desquamation from the SC starts to gradually slow. Cell turnover continues to decline.	Lipids start to decline. Barrier function reduces leading to greater trans epidermal water loss (TEWL), epidermis begins to thin as cell turnover continues to decline and the junction between the epidermis and dermis becomes flatter. Surface starts to become more uneven, fine lines and wrinkles begin to appear as structural integrity declines in dermis.	SC can begin to thicken as desquamation continues to decline. Epidermis continues to thin, melanocytes reduce but control of the ones that remain becomes disordered resulting in age spots and uneven colouring. Structural elements continue to decline, sagging occurs at jowls and eyelids, wrinkles deepen. Loss of fat cells in hypodermis further reduces volume and enhances sagging appearance.

Figure 5.4 Visible skin changes with increasing age.

deeply but causes burning (see Box 5.3). As a result, it is important to protect against all wavelengths of UV radiation. An Internet search for 'Mexican lorry driver's skin' will show dramatic photographs of a lorry driver who had spent his life behind the wheel of a lorry with the side of his face exposed to the sun, showing just how damaging UV radiation can be. Harmful, highly reactive molecules called free radicals can be induced in the skin by UV radiation and these can cause various changes in the chemical balance of the skin to occur. At worst these can lead to skin cancers, but most commonly these imbalances trigger other changes that can speed up the visible ageing processes at all levels. Protection with sunscreens has been shown to reduce and slow the visible signs of ageing. Even in climates where high UV radiation would not be expected it has been shown to have an effect, and many dermatologists recommend using SPF 30 products on a daily basis. More information on UV protection in cosmetic products can be found in Chapter 8.

Smoking tobacco has been shown to have a detrimental effect on the skin, again speeding up the natural processes and making them more visible earlier in life.

BOX 5.3 EFFECTS OF UVA AND UVB RADIATION ON SKIN

Historically, the effects of UV exposure on the skin were separated into two areas of the UV spectrum:

- UVA – causes ageing;
- UVB – causes sunburn.

This 'shorthand' helps us understand the main effects of UV radiation and helps identify products that protect against these important sources of skin damage. However, we now know that too much exposure to either type of UV radiation can bring risks of changes to DNA and, in serious cases, cancer. Tanning is a response to the damage from UVB exposure, but once again UVA radiation can also stimulate a tanning response to a lesser extent.

Pollution has more recently been identified as a possible cause for premature, visible skin ageing. Small particles of pollutants can be very reactive and may trigger similar changes that lead to damaging free radicals being formed. This is another example of where skincare can make a difference. The protective properties of skincare products can help prevent or minimize some of these extrinsic factors. See Chapter 8 to find out more about antioxidants.

As a final thought, the pattern of changes described here are those found in white, Caucasian subjects. People of colour, those with different ethnic backgrounds, will all experience changes with age, and the pattern of changes will vary. Those with Afro-Caribbean heritage may have a more ashy appearance as the skin surface ages; people with an Oriental heritage will be more likely to experience pigmentation changes, rather than lines and wrinkles, as the first signs of age.

5.2.2 Does Skin Vary from Individual to Individual?

The basic structure and composition of skin are common to all individuals but significant known differences exist. One of the most obvious differences is skin colour – a result of the number of melanocytes and the quantity and type of melanin produced in the epidermis. Differences in colour arose as humans evolved these differences as a natural protective mechanism against the damaging effects of UV radiation in different parts of the world. Blood vessels in the dermis and the quality of the dermis itself also influence the overall colour of the skin.

In the 1970s, a scale known as the Fitzpatrick scale after its inventor was developed to indicate different types of skin in relation to the responses to UV radiation (Figure 5.5). The scale is still used to this day to give an indication of skin tone and the reactivity to UV radiation. The descriptions focus on burning (blood vessel response) and tanning (melanocyte response).

Beyond colour, the differences between individuals' skin that cosmetic scientists need to be aware of come down to sebum levels, hydration and sensitivity. When selecting suitable skincare products, customers will tend to make their choices based on the type of skin they think they have in relation to these parameters. The classic skin types are dry, normal, oily,

The Science Behind Skin and Skincare Products 121

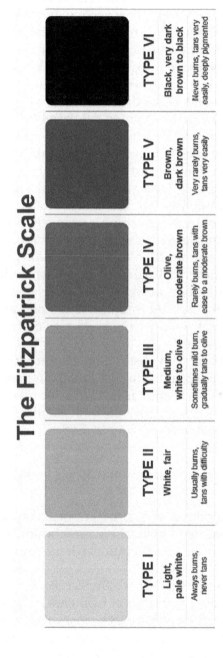

Figure 5.5 Fitzpatrick skin scale. Reproduced from ref. 1 with permission from Elsevier, Copyright 2017.

combination (patches of oily and dry) and sensitive. There are no strict definitions of these skin types and most people will self-diagnose, although not always very accurately.

Skincare products therefore need to be formulated and targeted at these different skin types, allowing a full choice for the consumer. Products aimed at users with dry skin will require a higher level of the key moisturizing components and usually a higher oil level. Conversely, products for users with oily skin will require less oils and are often lighter, gel-based formulas. This is important as oily and combination skins can lead to the individuals having a tendency towards spots and acne, which could be exacerbated by a rich, oily formulation.

Many people feel they have sensitive skins. This is particularly difficult to define but is often a result of feeling a slight reaction to heat, cold or even some previously used cosmetic products. Simple formulations with fewer ingredients and no fragrance are often the solution to this problem. There are also a number of soothing and calming ingredients than can be included in formulations designed for people who feel they have sensitive skin, *e.g.* α-bisabolol, allantoin and *Centella asiatica* (sometimes known as tiger grass). See Chapter 9 for more on 'sensitive skin'.

In many cases there is a distinction between the skin conditions described above and skin diseases. Skin products aimed at dry skin cannot treat eczema, products for oily skin cannot treat acne. These issues are covered by regulations in different parts of the world.

5.2.3 Are There Differences Between Men's and Women's Skin?

Although the components and structure of men's skin are the same as those of women's skin, there are hormonal differences in particular that need consideration when formulating skincare products.

Testosterone levels contribute to thicker skin in men, said to be around 25% thicker than that of women. Collagen density is also higher in men and tends not to be lost so dramatically from the age of around 50 years. During menopause, oestrogen reduction results in collagen loss, which in turn affects skin shape and structure. These differences are probably why men generally appear to age less quickly at this age.

Beard growth is probably the most visible difference between the facial skin of men and women. Men's stratum corneum is slightly thicker and this, coupled with the increase in active hair follicles, leads to a significantly rougher texture.

Higher testosterone levels are also responsible for more sebum in male skin. For women, changes in the balance of the female hormone oestrogen during the menstrual cycle can lead to variations in the oiliness of skin through the month. These hormonal effects can be particularly high at puberty and just after, which can lead to skin conditions related to overproduction of sebum such as acne.

Acne is characterized by inflammation of the sebaceous duct caused by the interplay of four key factors: hyperkeratinization (overproduction of keratin), excessive sebum secretion, colonization of *Cutibacterium acnes* (formerly *Propionibacterium acnes*) and inflammation.

Acne begins when the sebaceous ducts become plugged with corneocytes. Just as skin cells on the skin surface are constantly being sloughed off and renewed by the process of desquamation, corneocytes are also sloughed off the skin inside the pore. When sebum and corneocytes become trapped in the narrow opening of the pore, this can cause cells to clump and form a comedone. Comedones may be microscopic or visible to the eye as blackheads or whiteheads.

Sebum can build up and swell the follicle; *C. acnes* proliferate in this environment, triggering the body to respond by sending lymphocytes and phagocytes (cells that combat infection) to fight *C. acnes*, and consequently the skin becomes inflamed with pustules and papules. Figure 5.6 illustrates the mechanism of acne development.

- *Stage 1 – Microcomedone*
  Increased keratin production and excess sebum secretion cause corneocytes to stick together.
- *Stage 2 – Comedone*
  Shed corneocytes and sebum accumulate and form a plug that causes a closed comedo (whitehead) or open comedo (blackhead).

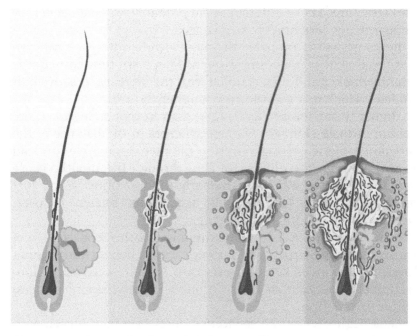

Figure 5.6 Pathogenesis of acne (© Shutterstock).

- *Stage 3 – Papule/pustule*
  Cutibacterium acnes proliferate, initiating an immune response characterized by mild inflammation.
- *Stage 4 – Cyst/nodule*
  Marked inflammatory responses and rupture of the follicular wall, which may lead to scarring.

5.3 STAYING ON THE SURFACE – DO COSMETIC INGREDIENTS GO INTO THE SKIN?

Cosmetic products, by definition, do most of their work on the surface of the skin. Moisturization, being the main function of skincare, has most effect on the stratum corneum, slowing the rate at which the cells dehydrate and shed, plumping the surface and maintaining the healthy appearance of this topmost layer of skin cells. Many myths exist on dealing with the real concern about if, and how deep, skincare products and their ingredients penetrate into the skin. The question of product safety is dealt with in Chapter 9 and myths are debunked in Chapter 10. However, it is worth covering the question here while skin function is still clear in our minds.

We have already discussed how the skin acts as a barrier preventing foreign materials from entering and there is evidence that certain molecules do have the ability to travel through the stratum corneum and epidermis. In rare cases, for example transdermal drug patches for substances such as nicotine, chemical compounds can reach the bloodstream. It is important, however, that we understand how difficult this journey actually is.

Figure 5.7 shows the stratum corneum in greater detail. Often described as a bricks and mortar structure, the stratum corneum walls are made up of specially toughened proteins coated with a lipid bilayer. This is a very good barrier and in most instances it does not let materials into the cell. An alternative path for any skincare molecules will be between the corneocytes, but this space is filled with a highly specialized multilayered lipid. This 'mortar' region repels most hydrophilic skincare ingredients.

Another route into the skin could be *via* the hair follicles. Travelling down the hair shaft could be a short cut to the

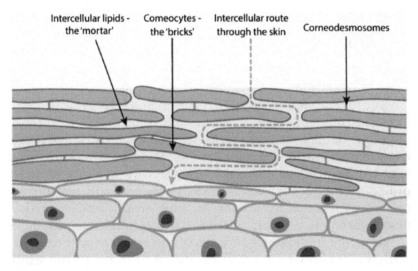

Figure 5.7 Pathway through the stratum corneum.

epidermis, However, the sebum produced in the sebaceous glands may still present a considerable barrier to this happening.

In order to reach lower levels of the skin, any ingredient molecule must travel through a complicated and long journey before it can potentially influence structural components such as collagen. Further constraints, such as the molecular size and polarity of the ingredient, make this journey potentially very difficult. For these reasons, most skincare ingredients remain on or in the stratum corneum. Having said this, some small molecules, such as peptides and retinoids, can, in the correct formulation, penetrate deeper into the skin, where they have the potential to bring about cosmetic changes. See Chapter 8 for more on ingredients.

It must be remembered here that a cosmetic product must not have a long-term physiological function on the skin or by definition it then becomes a drug, not a cosmetic. The associated regulations around drug development make this a very different discipline and cosmetic products must not cross this line.

5.4 WHAT GOES INTO SKINCARE PRODUCTS AND WHY – CARE AND PROTECTION FOR YOUR SKIN

If skin is so amazing, why would you need to put anything on it? There are many factors behind this question – some scientific, some cultural, some a combination of these and others. Here, we will try to stick to the science. What this tells us is that once you cleanse the skin, or expose it to adverse environmental conditions, the protective qualities of the skin begin to suffer in the long term. Loss of protective surface lipids and damage to important structural and defensive proteins deeper within the skin are just some of the factors that lead to the need to maintain and improve the skin's innate protective properties. Add to this the scientifically established role of appearance in communicating health and attractiveness, it becomes clear that care and protection of the skin have a major role in personal care. This is consistent with the definition of a cosmetic product as described in the Introduction to this book.

5.4.1 How Do Cosmetics Make a Difference to Skin Appearance?

This largely depends on the formulation and the composition of the raw materials, but it is fair to say that the changes in the skin's appearance after the application of a skincare product can be divided into immediate changes and longer term changes. By immediate we mean within a few minutes or hours of application, and by longer term we mean over several weeks of regular application and sometimes after several months.

To understand why the immediate changes occur, we need to understand a little about the optical properties of skin.

The stratum corneum and to some extent the epidermis and dermis are semi-transparent. Incident light reflects directly from the surface of the stratum corneum; the smoother the surface, the less scattered is the reflection. Owing to the semi-transparent properties, some of this incident light is absorbed, scattered and reflected by chromophores such as melanin, blood vessels and proteins in the deeper layers of the skin. This secondary, diffuse reflected light is what gives us the ability to see the multitude of different colours of skin (Figure 5.8).

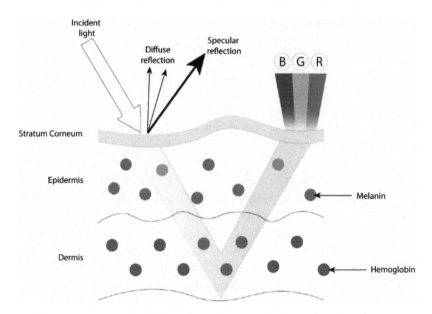

Figure 5.8 Diffuse reflection *versus* specular reflection on the skin's surface.

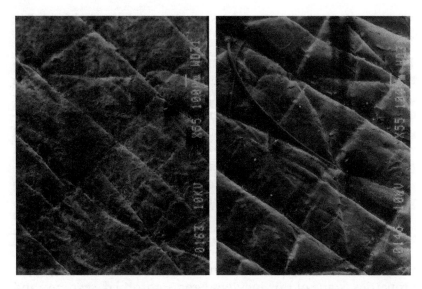

Figure 5.9 Scanning electron microscope images showing the difference between dry and moisturized skin surfaces. Left: dry skin – substantial diffuse reflection. Right: moisturized skin – smooth, reflective surface.

On a dry skin surface, the dead corneocytes build up and reduce the ability of the incident light to penetrate. More light is reflected and scattered, leading to a less translucent appearance. A well-moisturized, exfoliated surface appears smoother, brighter and ultimately younger. Even a very simple, well-formulated skin cream can have dramatic results. Figure 5.9 illustrates the surface of the skin taken with an electron microscope before and after application of a skincare moisturizer.

The longer term, visible changes that skincare products are capable of producing are far more complex and tend to be defined by the active ingredients used. We shall not describe these ingredients in detail here but, as used already as an example, peptides have been shown to have the ability to help the skin regain some of the structural integrity lost during ageing. This often manifests as a slight reduction in the appearance of lines and furrows, or more widely referred to as 'wrinkles', again resulting in smoother, more even, younger looking skin.

The effect of the pH of skin is outlined in Box 5.4.

> **BOX 5.4 PH OF SKIN**
>
> The very outer layer of the stratum corneum has a slightly acidic pH (4.5–6.5) and is therefore known as the acid mantle. The acidity is due to a combination of secretions from the sebaceous and sweat glands. The acid mantle functions to inhibit the growth of harmful bacteria and fungi. The acidity also helps maintain the hardness of keratin proteins, keeping them tightly bound together. When the pH of the acid mantle is disrupted (becomes alkaline), which can happen after using alkaline cosmetic products, particularly soaps, the skin becomes prone to infection, dehydration, roughness, irritation and noticeable flaking.

5.5 MOISTURIZATION – WHERE CHEMISTRY MEETS BIOLOGY

We have already established that to make fast, visible differences to the appearance of skin, the most important attribute for any cosmetic skin product is moisturization. Moisturizers usually take the form of an emulsion (think body butters, creams, lotions, milks), but are also available as oils, gels and serums, with new textures constantly being developed. Despite a huge choice of products available on the market, moisturizers usually contain very similar combinations of similar ingredients, most of which aim to moisturize and hydrate the stratum corneum.

5.5.1 What Is the Difference Between Moisturization and Hydration?

The terms moisturization and hydration are often used interchangeably and are associated with adding 'moisture' to skin to create a smoother, softer skin feel and appearance. Despite being undeniably linked, the definitions and also the ingredients used to maintain these two skin parameters are not the same.

Skin hydration relates to the concentration of water in the stratum corneum. Water in the epidermis originates in the cells at the lower epidermal layers. As these move upwards to become transformed into corneocytes, a water gradient develops from deep, water-rich layers to the less hydrated surface cells. To put this into context, the lower levels of the epidermis have a water content of around 70% and the outermost stratum corneum layer has an optimum hydration of around 20–30%. At levels below 10–20%, skin will appear dry and scaly and may begin to crack.

At the skin surface, water is lost to the environment through evaporation in a process known as trans-epidermal water loss (TEWL). The rate of TEWL and consequently the level of water in the skin are dependent on many factors, such as environmental humidity, skin barrier health and the water-holding capacity of the skin.

The water-holding capacity is assisted by natural moisturizing factors (NMFs) – the skin's built-in protector against dry skin (see Box 5.5). NMFs are contained within the corneocytes and increase the water-holding capacity of the cells, protecting the skin against cracking. Although vital in controlling the moisture levels in the skin, the NMFs can become depleted due to poor health or external factors, further exacerbating dry skin conditions. Maintaining optimum water levels in the SC is vital for creating smooth, supple skin.

Skin moisturization relates to smooth suppleness, which is further assisted by skin 'oils'. The lipidic material in the skin's surface structure found between skin cells – described earlier as the 'mortar' in the bricks and mortar model – comprises cholesterol, ceramides and free fatty acids arranged in a bilayer structure between corneocytes. These highly organized structures form an impermeable barrier that gives integrity and helps reduce water loss. Barrier function is a term used in skincare and describes the effectiveness of this lipid barrier. Over-cleansing, water exposure and extremes of temperature and humidity can impair this barrier, increasing

---

BOX 5.5 WHAT IS AN NMF?

An NMF is a mixture of materials such as amino acids, urea, mineral salts and organic acids – the breakdown products of epidermal cell components in the stratum corneum. These hygroscopic 'water magnets' attract and hold water in the stratum corneum.

TEWL. An impaired barrier can also facilitate the penetration of potential irritants and allergens, which may disrupt the barrier further.

Adding water alone to dry skin is futile if the barrier function is not being maintained as water is lost as quickly as it is added. Dehydrated skin does, of course, require treatment with hydrating ingredients, but moisturization through oil addition is equally important.

The important thing for a cosmetic formulator to remember is that oil and water components have equal importance – something we shall come to shortly.

5.5.2 Which Ingredients Are Important in an Effective Skin Moisturizer?

Well-formulated moisturizers usually contain a combination of hydrating and moisturizing ingredients working together to optimize the water and lipid content in the SC through the different mechanisms described.

Humectants: These are hygroscopic molecules that attract, absorb and hold onto water from the atmosphere; when added to a cosmetic formula, humectants simulate the work of the NMF. Common examples include glycerine, sorbitol, urea, propylene glycol and pyrrolidonecarboxylic acid (PCA). Hyaluronic acid, a glycoprotein found naturally in the dermis but made using biotechnology, has also become common in modern skincare formulations.

Emollients: Defined as 'softening', these are usually oil-soluble ingredients derived from a wide range of synthetic and natural sources. Some emollients also have an occlusive action – preventing water loss, thus acting in two ways – smoothing the surface **and** reducing TEWL. Many emollient ingredients have a similar chemical composition to that of natural skin lipids – for example, squalane is closely related to a constituent of sebum (see Box 5.6); lanolin, or wool fat, and its derivatives also have a similar composition to sebum and are excellent skin moisturizers. Cocoa butter has high levels of palmitic acid, evening primrose oil is rich in linoleic acid – these fatty acids are important components of SC lipids.

> **BOX 5.6 SQUALANE *VERSUS* SQUALENE?**
>
> These two ingredients sound very similar, but they are different.
>
> *Squalene* is a natural organic compound found in skin sebum. As a cosmetic ingredient it was originally extracted from shark's livers (hence the name – *Squalus* is a genus of shark). Thankfully, it is now vegetable derived, usually from olive oil.
>
> *Squalane* is produced by the hydrogenation of squalene and is a useful cosmetic emollient.
>
> Squalane is more stable than squalene.

Some emollients are used for their light feel on the skin – esters such as ethylhexyl palmitate and isopropyl palmitate and fatty alcohols such as octyldodecanol and caprylic/capric triglyceride. Heavier, more occlusive emollients include vegetable oils, examples being sweet almond oil, coconut oil and jojoba oil. Natural butters – shea butter and cocoa butter – melt at around skin temperature, offering intermediate sensory properties.

Petroleum-derived mineral oils and petroleum jelly (Vaseline®) literally 'lock' moisture into the skin and tend to be found in more medicinal moisturizers aimed at serious dry skin conditions.

Hydrating ingredients: Many natural, plant-derived ingredients have been shown to have superb hydrating properties, superior to that of water alone. Aloe vera, for example, although consisting of 98% water, also contains a complex mixture of minerals, vitamins, sugars and salts that are well known to moisturize skin.

As with all cosmetic science, the skill comes in combining materials from all these groups of ingredients to produce an effective, long-lasting formula that balances the needs of dry skin while also feeling pleasant to use. We will learn more about how to combine them successfully in Section 5.6 on emulsions and emulsifiers.

5.6 EMULSIONS – BETTER TOGETHER!

Hopefully you haven't forgotten that the skin performs its main functions by holding onto water by the combined action of oils and humectants. So it should be no surprise that typical skincare formulation ingredients usually fall into two categories – water soluble/compatible (hydrophilic) and oil soluble/compatible (lipophilic).

The humectants discussed earlier are usually water soluble and need to be combined in a water phase. The emollients are oil soluble or dispersible and likewise need to be together in an oil phase. We already know that the two types of materials are vital for the success of the final formula, so it is critical that the two phases are made to work together in a stable formulation. Oil and water do not mix and no matter how much energy is put into trying to make a homogeneous product they will quickly separate when left to stand. For this reason, most cosmetic skincare products are emulsions. Emulsions are formed when an oil phase is combined with a water phase in the presence of an emulsifier.

Emulsifiers are the cosmetic chemist's best friend! Figure 5.10 shows the principle of emulsifier action. They are molecules containing both a hydrophilic and a lipophilic element – part water (hydro) loving, part oil (lipo) loving, and often symbolized as a tadpole-shaped molecule with a hydrophilic head and lipophilic tail. These molecules can span the interface between the water and oil to keep the dispersed phase suspended. Discrete droplets of the dispersed phase will form, protected by the emulsifier shell. Emulsions can be either oil-in-water (o/w) or water-in-oil (w/o) depending on the emulsifier used and the proportion and chemistry of the two phases (see Box 5.7). For reasons that it is hoped are obvious, the two phases are also called the *internal* (dispersed) and the *external* (continuous) phase.

BOX 5.7 WHAT TO CONSIDER WHEN CHOOSING AN EMULSIFIER?

The selection of the emulsifier for any given system depends largely on the oils used and on the desired properties of the final product. Oil-in-water emulsions are the most popular for most skin creams as they tend to give lighter, less oily feeling products. However, this can vary regionally and in some markets water-in-oil creams may be preferred.

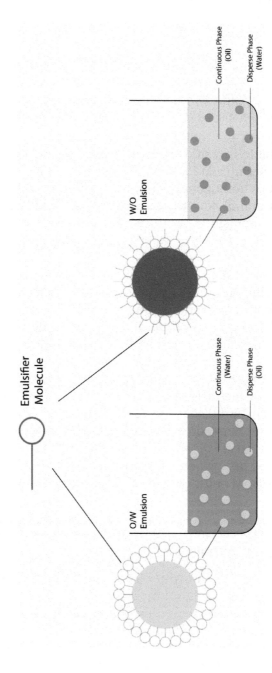

Figure 5.10 Emulsifier arrangement in an oil-in-water emulsion (left) and a water-in-oil emulsion (right).

5.6.1 What Are Emulsifiers and Why Are They All Different?

Chapter 2 considered surfactants – a general term for surface-active agents – and how they allow oils and water to interact. The term 'emulsifiers' can be thought of as a subgroup of surfactants used to produce emulsions. A system for classifying emulsifiers was developed and published in 1949 and assigns an HLB (hydrophilic–lipophilic balance) value to any given emulsifier. This can be used to identify the correct emulsifier needed for any particular combination of oils and aqueous ingredients. The classic diagram in Figure 5.11 shows how the HLB value relates to the properties of the emulsifier and how it can be used to select the correct emulsifier for the formulation required.

The HLB system was designed to work with ethoxylated, non-ionic emulsifiers, although in theory it can be applied with limitations to any emulsifier. The science behind HLB calculations can be extremely complicated and will not be covered here,

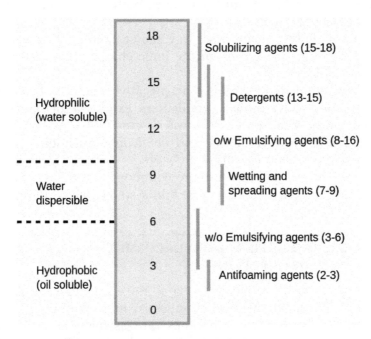

Figure 5.11 HLB values and properties of different emulsifiers.
Reproduced from https://commons.wikimedia.org/wiki/File:HLB_scale.svg, under the terms of a CC BY-SA 3.0 license, https://creativecommons.org/licenses/by-sa/3.0/deed.en.

but the HLB system is still used today as a useful guide when looking at the various emulsifier options available. However, since its introduction, the science of emulsions and emulsifiers has developed enormously and a vast array of sophisticated and very versatile emulsifiers are now available to the cosmetic chemist, making the HLB system a tool that is not necessarily applied to every formulation (see Box 5.8).

The modern cosmetic chemist needs to understand the different types of emulsifiers and the basic chemistry involved in their manufacture and structure. This is important as they define which ingredients they will work with and the final stability of the formulation. As the move towards increasingly natural and sustainable formulas continues, understanding the chemistry is a vital factor in making any ingredient selection. Most modern emulsifiers will start life as a vegetable oil, often palm oil, or a synthetic mineral-based oil. Industrial chemistry is then used to convert these oils into far more complex molecules that exhibit the ability to have both hydrophilic and lipophilic character.

Like the detergents outlined in Chapter 2, emulsifiers can be broadly grouped according to the ionic charge associated with the molecule.

Anionic emulsifiers exhibit a net negative charge at the head group; conversely, *cationic* emulsifiers exhibit a net positive charge at the head group. *Amphoteric* emulsifiers exhibit both positive and negative charge and *non-ionic* emulsifiers, as the name suggests, do not exhibit a charge.

Some of these groups are more common than others, but all have specific benefits and suit certain systems. Many traditional

---

BOX 5.8 EMULSIFIERS USED IN COSMETIC FORMULATIONS

The range of emulsifiers available to the cosmetic formulator is vast and many chemical suppliers manufacture their own blends of emulsifiers, sometimes with as many as three or four different emulsifying materials, all with different HLB values. These all have different benefits and are tailored towards specific formulation types.

skincare creams used the emulsifying power of simple soaps produced in a reaction between fatty acids, for example stearic acid, and an alkali such as sodium hydroxide during the mixing process. The resulting stearate soap (sodium stearate) acts as an effective anionic emulsifier. The downside of these systems was the fact that, even when weaker alkalis such as triethanolamine were used, the resulting pH was relatively high and not ideal for skin preparations.

Non-ionic emulsifiers became far more popular in skincare products and fatty alcohol ethoxylates are probably the most common. The polysorbate series of emulsifiers are the best known but there are many hundreds of commercially available non-ionic emulsifiers. Any material with the acronym PEG-[poly(ethylene glycol)] in the INCI (International Nomenclature of Cosmetic Ingredients) name is likely to exhibit non-ionic emulsifying properties. These emulsifiers are very effective and well tolerated by skin as they work at relatively low use levels. With the growth of 'green chemistry', and the continuing move towards 'natural' skincare, PEG-based, non-ionic emulsifiers are increasingly less popular. Alternative non-ionic emulsifiers, for example cetearyl glucoside, produced by etherification of fatty alcohol with glucose, are now widely used. Polyglyceryl esters are now also available as non-ionic PEG alternatives.

Cationic emulsifiers are less common in skincare formulas but can be useful for producing different textured emulsions with particularly smooth skin feel. More commonly used as conditioning ingredients in hair care, cationic molecules can be multifunctional. However, they can be difficult to use and careful selection of the oil and water components is critical.

Amphoteric emulsifiers are more often used as surfactants in shampoos and body washes; they do not make particularly useful emulsifiers in skincare products as they tend to change their ionic character depending on the pH of the system – at high pH they tend to be anionic and at low pH they exhibit cationic characteristics. A common example of an amphoteric molecule is cocamidopropyl betaine.

Skincare formulations have been transformed over the last 20 years by the introduction of synthetic silicone fluids. These impart a wonderfully smooth and lubricious skin feel and are discussed in more detail in the rheology section (see Section 5.7).

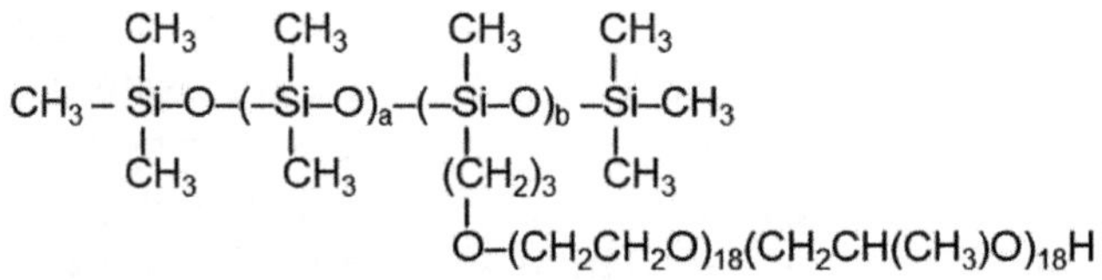

$$\begin{array}{c} CH_3 CH_3 CH_3 CH_3 \\ CH_3-Si-O-(-Si-O)_a-(-Si-O)_b-Si-CH_3 \\ CH_3 CH_3 (CH_2)_3 CH_3 \\ O-(CH_2CH_2O)_{18}(CH_2CH(CH_3)O)_{18}H \end{array}$$

Figure 5.12 Structure of PEG/PPG-18/18 dimethicone. Reproduced from ref. 2 with permission from Personal Care Magazine.

Formulating with silicone fluids can be difficult, however, and emulsions can be particularly problematic. To solve this problem, the manufacturers of silicone ingredients developed specialist silicone emulsifiers based on non-ionic PEG systems but incorporating a silicone element. These will often appear as PEG/PPG-X/Y dimethicones in their INCI descriptions, where X and Y represent the number of repeating units of PEG and PPG [poly(propylene glygol)], respectively. These silicone emulsifiers have become important and useful ingredients for the modern cosmetic scientist. An example of these is PEG/PPG-18/18 dimethicone; the complicated chemical structure of this molecule is shown in Figure 5.12.

A quick survey of well-known cosmetic emulsion formulas produced the list of popular emulsifiers in Table 5.1, illustrating the groups of emulsifiers now available.

5.6.2 How Can We Make Emulsions Stable for Several Years?

Even with an effective emulsifier, emulsions are dynamic systems and will eventually separate out into their component oil and water phases. For a commercial cosmetic emulsion to be of any practical use, it needs to remain stable for several years under normal conditions, so it is important to understand how to achieve this.

Rarely does a formula contain just one emulsifier. Co-emulsifiers are secondary ingredients that are chosen to help support the primary emulsifier in any system. As with many systems, multiple combinations can be more effective than the sum of their parts, and this is often true of emulsifier blends.

Table 5.1 Example of emulsifiers used in cosmetic products and their properties.

INCI name	Category	HLB value	Comments
Glyceryl oleate	Non-ionic	3	w/o emulsifier
PEG/PPG-19/19 dimethicone	Non-ionic	3.5	w/o but specifically w/si (water-in-silicone) emulsifier
Sorbitan oleate	Non-ionic	4.3	w/o emulsifier
Sorbitan stearate	Non-ionic	4.7	w/o emulsifier
Steareth-2	Non-ionic	6	o/w emulsifier usually combined with steareth-21, ethoxylated
Sucrose laurate	Non-ionic	11	o/w emulsifier
Distearyldimonium chloride	Cationic	11	Usually used as a hair conditioner. (See Chapter 3 – Good Hair Day)
Glyceryl stearate citrate	Anionic	12	o/w emulsifier
Cetearyl glucoside	Non-ionic	13	o/w emulsifier
Palmitamidopropyltrimonium chloride	Cationic	14	Usually used as a hair conditioner
Polysorbate 60	Non-ionic	15	o/w emulsifier, ethoxylated.
Cetearyl alcohol	Non-ionic	15.5	Weak o/w emulsifier. Important co-emulsifier
Steareth-21	Non-ionic	16	o/w emulsifier, ethoxylated
Glyceryl stearate (and) PEG-100 stearate	Non-ionic	19	o/w emulsifier, ethoxylated
Sodium stearoyl glutamate	Anionic	23	o/w emulsifier, non-ethoxylated

Secondary emulsifiers will have slightly different HLB values to the primary emulsifier and will help broaden the effectiveness of the emulsifying system. A common secondary emulsifier used in o/w systems, for example, is cetearyl alcohol. This combination of fatty alcohols has weak emulsifying properties but can form a lamellar gel structure between the oil and water phases, helping to stabilize the system.

Figure 5.13 shows a representation of different emulsifiers with respect to their shape, size and hydrophobicity (water/oil

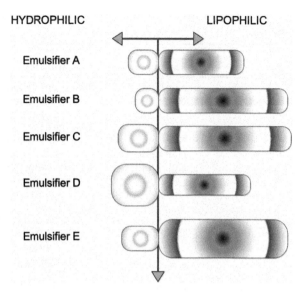

Figure 5.13 Emulsifier structure based on hydrophilicity and hydrophobicity. Reproduced from ref. 3 with permission from Springer Nature, Copyright 2012.

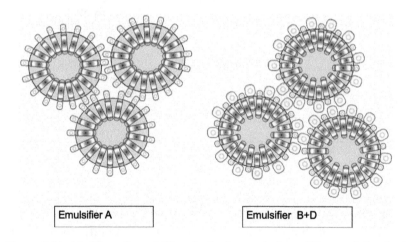

Figure 5.14 Packing of emulsifiers in the internal phase (dispersed droplets) of emulsion.
Reproduced from ref. 3 with permission from Springer Nature, Copyright 2012.

compatibility). By mixing emulsifiers, more effective, closer packed shells can be created at the oil/water interfaces, as illustrated in Figure 5.14 (see also Box 5.9).

> **BOX 5.9   MIXING EMULSIFIERS? NOT ALL EMULSIFIERS ARE COMPATIBLE WITH EACH OTHER**
>
> A major word of warning is important here. Not all emulsifiers can be mixed and used in the same system. Because of the charge on ionic emulsifiers, anionic and cationic emulsifiers cannot be used together. The strong interactions that occur between them due to their opposite charges will result in unstable emulsions. Care must also be taken to avoid other ionic components of the formula when using them. Non-ionic emulsifiers are generally much more widely compatible with other cosmetic ingredients.

Alongside secondary emulsifiers, probably the most important stabilizing ingredients are suspending agents and thickening ingredients. These help to suspend the emulsion droplets, preventing them from coalescing and leading to instability. We talk more about these in the rheology section (see Section 5.7), but they are a critical part of the emulsion system as a whole and often hold the key to producing emulsions that remain stable for long periods of time under varying conditions of temperature.

5.7   TOUCH AND TEXTURE – IT'S JUST A FEELING

We're told not to judge books by their covers but, like it or not, first impressions are influential and it's no different in skincare. When a cream, gel, lotion or serum is first applied to the skin, the user makes an instant decision about how it feels. As we learned at the beginning of this chapter, skin is a sensory organ and touch is a very important skin function. The way a product spreads, the way the texture changes as it goes onto the skin, the temperature sensation, the smell, the feel on the skin as it dries and any residual tackiness or shine after it is applied – all of these things matter.

Our brains interpret these skin sensations and, although the user has no idea whether the effects promised on the product pack will be realized, the initial sensations will play a major role in deciding if it's the right product for them. So, just like the illustrator of a book cover, the cosmetic scientist needs to get all these textural attributes right if the formula is to be successful.

5.7.1 What Is Rheology and Why Is It So Important in Skincare?

The science behind these sensorial properties and the secret to controlling them lie in rheology. Rheology is defined as the study of the flow of matter, usually in liquid form but also in soft solids, such as skin creams. The full scientific definition relates the shear stress applied to the liquid to the shear rate and the effect on the resulting apparent viscosity of the fluid. It is classically described as illustrated in Figure 5.15.

A Newtonian fluid such as water or ethanol will maintain the same viscosity when shear is applied to it; this is relatively rare and most fluids, certainly the skincare products in which we are interested, will not and are therefore described as showing non-Newtonian behaviour. Shear thickening and shear thinning are the two most important properties for us to consider. Classic examples of these properties from everyday life are often used, the most common example of shear thinning being ketchup, which thins when shaken, returning to its original viscosity

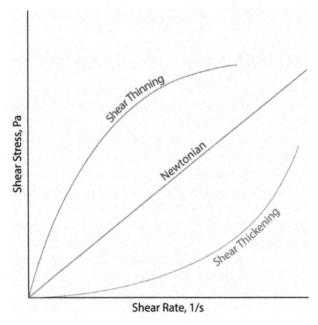

Figure 5.15 Graph showing changes in shear stress as a result of shear rate for Newtonian and non-Newtonian fluids.

when left to rest. Conversely, and more unusually, a mixture of cornflour in water will thicken when stirred, thinning back to its original viscosity when mixing is stopped, is an example of shear thickening. We are mostly interested in shear thinning and the way to control different shapes of this particular curve in the graph (Figure 5.16). A graph relating shear rate to viscosity is a simpler way to think about and understand shear thinning and shear thickening.

Rheology as a study can become quite complex and, in order to understand it fully for an individual cosmetic formulation, the chemist needs to make various measurements on a rheometer. In practical terms, most formulators will only have viscometers in their laboratories, which will measure viscosity – how 'thick' or 'thin' a sample is. The rheology of the formulation tends to be assessed by the formulator subjectively through touch and experience, in exactly the same way that the end user assesses rheology without realizing it.

Two different skin creams can have the same apparent viscosity, and they can appear very similar in a jar or bottle.

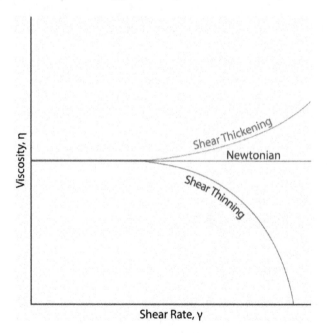

Figure 5.16 Graph showing changes in viscosity as a result of shear rate for Newtonian and non-Newtonian fluids.

However, as soon as they are subjected to shear stress, *e.g.* on removal from the jar with a finger, they start to feel different. As they are applied to the skin these differences can be stark and all depend on the degree of shear thinning experienced. The cream with a higher tendency to shear thin will quickly break and feel almost 'watery' as it spreads and dries. The cream with a lower tendency to shear thin will remain thick and 'heavy' as it is applied, taking longer to spread and longer to dry and ultimately leaving a more obvious film on the skin. Neither of these attributes is right or wrong, they are just different. The cosmetic scientist must balance these properties to achieve the result required.

5.7.2 How Can We Control the Rheology and Skin Feel of Cosmetic Creams?

The thickening system selected for any given cosmetic formulation probably has the most influence on the final rheology. Thickening ingredients work alongside the emulsifiers in any emulsion to aid with viscosity control and, critically, to improve the stability of the formula. The challenge for any formulator of cosmetic skincare products is to balance the level and combination of these different thickeners. Rarely can the right texture be achieved by the use of one thickener alone. A trial-and-error approach is often the best way to achieve the desired rheology and most formulations on the market today contain complex mixtures of the thickening ingredients described here. Often these components are used at surprisingly low concentrations, as subtle additions can have important effects on the perceived final skin feel.

The different types of thickeners can be broadly broken down into four groups, as follows.

5.7.2.1 Vegetable Gums. Vegetable gums are widely used to help build viscosity and to improve the stability of emulsions. They all exhibit slightly different properties and all have pros and cons. Although excellent at both viscosity control and stability improvement, gums tend to be sticky when they dry down and, if used in too high quantities, can feel 'slimy' when applied to the skin. For this reason, they have to be used at relatively low levels and almost always alongside other thickening systems.

Typical examples of vegetable gums are guar gum, xanthan gum, locust bean gum and carrageenan gum.

The natural origin of these gums has given them a resurgence in recent years with the move back towards more natural, less synthetic formulas. It is always interesting to look back at the history of any subject and, in cosmetic science, gums have always played a leading role. In the well-known book on cosmetic science by De Navarre published in 1941, gums are described as mucilage.[4] A popular one at the time was quince seed mucilage, which required relatively complicated processing methods in comparison with the gums available today.

5.7.2.2 Polymers. It is fair to say that the invention of carbomers in the 1950s revolutionized cosmetic science. Carbomer is a trade name for a group of polymers more correctly described as polyacrylic acids. In the acid state these synthetic polymers disperse into water-based systems to give milky, low-pH, water-thin liquids. When neutralized to pH 5 and above, the liquids transform to clear gels; the higher the concentration, the thicker is the gel. In skincare emulsions these neutralized gels help to emulsify, thicken and control the final skin feel, making them truly multifunctional ingredients and, unlike gums, they do not leave a sticky after-feel on the skin. Since their introduction, many variations on the chemistry have been introduced that give subtle differences with respect to ease of use and performance. They remain one of the most important and versatile ingredients available to the cosmetic scientist today.

Related to carbomers are a whole class of similar acrylate-based polymers. Sodium polyacrylate is an example of a pre-neutralized polymer that exhibits high water absorbency and works well as a thickener at low concentrations in skincare formulations – again, helping to control the rheology of an emulsion or gel.

5.7.2.3 Starches. Naturally derived ingredients have always been important in cosmetic science and their popularity has seen a huge resurgence in recent years. Starches available from vegetable sources such as rice, potato, tapioca and corn can contribute interesting rheological properties to skincare formulations. They often help to thicken formulas as they swell in aqueous environments, but owing to their insolubility and very fine grain size, when dry they can contribute a smoothing, silky after-feel to many products. Care must be taken not to overload formulations with starches as they can clump and roll up on the skin during application and dry down. This is another example of how the formulator's skill balances the different ingredients available for rheological control.

5.7.2.4 Waxes. Waxes are defined as lipophilic, high molecular weight, mouldable solids that encompass a very broad group of both natural and synthetic ingredients. Beeswax is probably the best-known example of a wax. By incorporating beeswax into a cosmetic emulsion, the viscosity will be increased and the rheology modified. Beeswax tends to make creams feel heavy and less shear thinning, often useful for richer formulations that are designed to moisturize deeply and remain for longer on the skin surface. Cetearyl alcohol is a very common example of a fatty alcohol used widely in skincare formulas and many other types of personal care formulations. This also can be loosely described as a wax. Fatty alcohols can form lamellar gels, which are crystalline, colloidal structures that thicken water-based formulations and dry down to exhibit a soft, almost powdery after-feel. High molecular weight esters such as myristyl myristate can also be described as waxes and again can contribute really useful properties to finished skincare formulations; this particular wax imparts a smooth, silky skin feel to cosmetic emulsions. These are a useful group of ingredients to consider when designing the overall feel and rheology of a skincare formulation.

5.7.3 By What Other Ways Can We Affect the Feel of Skincare Formulas?

The oil phase of a skincare formulation is an important factor in how the product will feel during and after application. The word 'oil' is used to describe the lipophilic, liquid components of the formulation. These are generally either vegetable oils or

> **BOX 5.10   OILS WITH DIFFERENT PROPERTIES**
>
> Castor oil, for example, is a relatively thick, heavy vegetable oil used widely in cosmetics, especially lipsticks. Sweet almond oil, on the other hand, is a light, low-viscosity oil that feels much drier and spreads much further when applied to the skin, therefore making it a great carrier oil for products such as massage oils and facial oils.

synthetic esters or combinations of both. Oils themselves are generally Newtonian fluids but they contribute to the overall rheology of the formula owing to their wide range of viscosities (see Box 5.10). Hundreds of vegetable oils are available to the cosmetic chemist and they all have unique fatty acid profiles that tend to define how they will feel. Again, combinations of different oils must be tried in order to find the perfect balance for any given formulation.

When describing cosmetic oils, it is common to hear the term 'dry oil'. This may seem like an oddity as we think of dry being the opposite of wet and pertaining to water. In the case of oils, it means that the end skin feel is non-greasy, non-tacky, silky smooth as if the oil has disappeared. Light, low-viscosity oils often exhibit this property and it is important when selecting the final combination for a skin cream. The balance needs to be struck between leaving an oily 'moisturizing' film and a smooth almost dry after-feel.

Silicone technology has arguably been one of the most important introductions to cosmetic science in the last 25 years. Most cosmetic 'oils', whether natural vegetable oils or synthetic esters, are based on organic, carbon chemistry – long-chain molecules with a characteristic carbon backbone.

In synthetic silicone technology, the backbone consists of silicon–oxygen repeating units. The resulting silicone polymers are known as siloxanes. The simplest cosmetic siloxanes are known as dimethicone (polydimethylsiloxane), with relatively short chains and simple side chains. Many more complex silicone fluids have been developed, all with differing properties, and possibly the best-known cosmetic silicone is cyclopentasiloxane, which as the name suggests has a cyclic structure and has a degree of volatility (Figure 5.17).

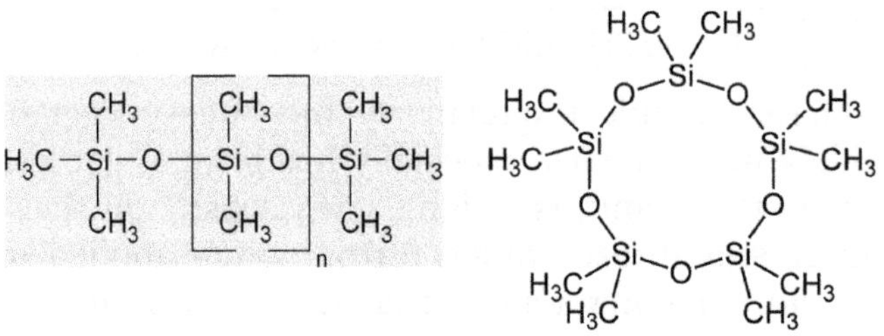

Figure 5.17 Structures of two types of silicones: left, dimethicone; right, decamethylcyclopentasiloxane.

The reason why silicone fluids have become such an important tool in cosmetic formulation is due to their great lubricity. They impart exceptional slip and smoothness to the skin feel of cosmetic creams, outperforming traditional 'oils' by some margin. When more complex silicone fluids, known as silicone elastomers, were introduced for the first time into cosmetic skincare, unique textures were achievable that had not previously been possible. This paved the way for a vast number of cosmetic silicones to be developed, all with slightly different properties.

Silicone fluids exhibit strong hydrophobic film-forming properties, which makes them excellent for protective skin barrier products and enhances waterproofing properties where needed. See Chapter 10 for more on silicones.

5.8 DIFFERENT TYPES OF SKINCARE PRODUCTS

We will now move on to discuss the different types of products available on the market, and their properties.

5.8.1 Why Do We Have Day and Night Moisturizers, and Are They Different?

Skincare products have long been formulated and marketed specifically for 'day' or 'night' use. In addition to moisturization, day creams offer additional protection from environmental stressors to which skin is exposed during the day, particularly UV radiation and pollution. The role of UV absorbers and sun protection factor (SPF) products and antioxidant ingredients designed to protect the skin from oxidative stress is covered in Chapter 8. Conversely, night creams may contain ingredients that are either unstable in sunlight or make the skin more susceptible to sunburn. α-Hydroxy acids (AHAs) and retinol are two examples – these need to be avoided in day creams where the user could be exposed to strong sunlight.

Day creams may also be lighter in texture, quicker drying, more easily absorbed and, importantly, more suitable to wear under makeup.

Night creams logically do not require an SPF and are designed to 'repair' skin from the damage during the day. Night creams traditionally contain more occlusive materials and display a heavier texture, providing more intense moisturization. This is considered more acceptable at night where skin feel and make-up application are not important considerations. The natural skin repair mechanisms occur at night – night creams may contain ingredients to help facilitate this repair and renewal.

5.8.2 Do Men Need Different Moisturizers to Women?

We have already discussed the differences between men's and women's skin, but what does this mean for the formulator? As with many areas of life, the gender gap has narrowed in recent years and, in reality, products aimed at men are not significantly different to women's products. However, subtle difference in texture and skin feel should be considered. An aftershave balm, for example, is usually a light, relatively thin, moisturizing emulsion containing soothing and often cooling active ingredients. How the product feels and is perceived by the user are always the primary consideration when formulating. If little attention is given to this aspect, then a product is sure to fail in the marketplace. For this reason, the most important element to developing any skincare product, be it for men or women, is to test it thoroughly on the target audience!

5.8.3 How Important Are Skincare Regimes and What Are the Necessary Products?

Cleanse, tone, moisturize has traditionally been the skincare daily mantra. Skincare regimes were born out of the relative crudeness of many early formulations from the past. Cleansers tended to remove some of the protective proteins and lipids from the skin, or leave a deposit on the skin; either way, the skin did not feel clean. Toners helped resolve these issues and left the skin cleaner and prepared for moisturization. However, many toners used alcohol to be effective, and this in turn needed better moisturizers to be effective. Modern-day skincare products have overcome many of these difficulties, although many users still feel the need for the three steps and, in some cases, even more.

Cleansing skin is essential to good skin health and remains one of the most important steps in any skincare regime. Cleansing products can be further broken down into groups:

- foaming washes used with water and then rinsed from the skin;
- creams massaged onto the skin and wiped off, removing dirt and makeup at the same time.

Hybrids of these two also exist and a more recent addition to the group of cleansers are products that effectively cleanse and moisturize. These are massaged onto the skin, lifting all dirt and makeup, and then wiped off with a hot, damp cloth. Usually balms, oils or wax-based formulas that melt at around skin temperature are used as they leave a protective layer on the skin.

Toners are the one group of products that are not universally applied by skincare users. Water-thin, aqueous liquids, toners traditionally contained alcohol or witch hazel, which acted as an astringent, and can help reduce the appearance of enlarged pores. More recently, they have been formulated with skincare active ingredients that enhance the cleansing process while adding additional moisture. α-Hydroxy acids (see Chapter 8) are often included in this type of product to further smooth and prepare the skin for the moisturizing step to follow.

Exfoliators are rarely used daily, but offer more thorough cleansing to remove stubborn dead skin cells in dry skin, or skin with problematic pores or blackheads. The exfoliation effect can be achieved by either abrasive scrub particles and/or mild acids on the skin and leads to noticeably smoother skin after use. The size and shape of the scrub particles vary, which affects the scrub efficiency. This is important; the formulator can control the level of exfoliation to avoid creating aggressive scrubs that can lead to sore, irritated skin.

The scrub particles are usually natural particles of ground nut shells, silica particles or microcellulose-based materials. Plastic microbeads were commonly used for this purpose until more became understood about their environmental fate. These were banned in cosmetic scrubs by most global authorities from 2017. See Chapter 10 for further information.

Serums have become an increasingly important category in the facial skincare market. Although traditionally a medical term referring to biological fluids, the word 'serum' comes from the Latin word *serum* meaning watery fluid or whey. Serum has been adopted by the cosmetics industry to describe concentrated, usually water-based, skincare formulas with high levels of active ingredients. This may be something that is harder to achieve in a conventional emulsion-based moisturizer and, as a result, they are often simpler formulations.

A good example of a serum is a product using vitamin C to even out uneven pigmentation. It could be targeted for use on fine lines around the eye area, for example, and could be used in conjunction with a conventional moisturizer. The moisturizing action of the serum in this case may be secondary to its efficacy. Alternatively, a serum could be based on a powerful moisturizing ingredient such as hyaluronic acid. In this case, the performance may well be better than that of a conventional emulsion-based moisturizer.

Facial oils – blends of anhydrous oils – either natural or synthetic or blends of the two – are another popular treatment used within skincare routines. These are commonly used in professional skincare where the user will be spending a prolonged period on the routine which is applied by a trained skin or beauty therapist. Many brands exist for home use, however, and many people like to use oils in place of more traditional night creams.

When formulating for facial oils, the chemist is limited to oil-soluble ingredients. Many if not most of the well-known skincare active ingredients available are water soluble, so the benefits of facial oils tend to be limited to moisturization by preventing TEWL.

Treatment masks are not new to the cosmetics industry but they have seen a revival over the last decade. Generally designed to be used a few times a week, depending on the type of formulation, the benefits of the mask can be extra moisturizing, soothing and calming, cleansing or pore refining. The functional base of formulation varies. Clays have been used of centuries, their ability to draw oil and dirt from the skin and silky residue being the main advantages. In some mask formulations, a pleasant warming sensation is produced on the skin when water is added thanks to certain ingredients such as zeolite and calcium chloride. On the other hand, the addition of cooling ingredients, for example menthol derivatives or alcohol, provides a refreshing, cooling effect on the skin which is particularly appealing for warmer climates.

Sheet masks are now a well-known format for this type of skin treatment. Popular in Eastern skincare, these consist of a substrate – usually cotton or another natural fibre – soaked in an active formulation base, usually contained in a sealed sachet. The mask is removed and placed over the face for a prescribed amount of time.

5.8.4 Are the Skin Concerns for the Body the Same As Those for the Face? – How Many Different Moisturizers Do We Need?

As we have seen already, although the composition and structure of the skin remain broadly the same all over the body, there are

Table 5.2 Properties of formulations for different parts of the body.

Product type	Key features
Eye creams	• Normally lighter formulations compared with face and body creams • Active ingredients target fine lines, wrinkles, dark circles and eye bags • Fragrance free to reduce potential irritation potential to delicate eye area
Hand creams	• Richer, thicker formulations than face creams. Tend to have a drier, less greasy after-feel due to higher wax content • UV absorbers often included to help reduce signs of ageing (*e.g.* pigmentation) caused by UV radiation
Neck creams	• Very similar to face creams. Active ingredients tend to target sagging skin and 'crepy' neck skin • Lifting and firming claims are often made for these products but sometimes can be difficult to substantiate
Foot creams	• Similar formulations to hand creams • Active ingredients target cracked skin on heels • High levels of urea tend to be used, *e.g.* 10% in a moisturizing cream, effective and rapid moisturization or at higher levels (of 20% and more), to give a keratolytic effect (helps the skin's natural shedding process) • High levels of urea can sometimes cause instability in formulations such as the development of ammonia odour and emulsion separation

differences in thickness, texture and hair follicle density. By and large, formulations are similar for the different body areas; effective moisturization tends to be the primary function.

There are, however, a number of exceptions to this where specific body concerns are targeted by cosmetic products (Table 5.2). Examples are cellulite, stretch marks and keratosis pilaris (small lumps on the backs of the arms). In each case, the products will contain specific ingredients designed to improve the appearance of these skin concerns with regular use. Examples of this are the use of lactic or glycolic acid in keratosis pilaris products and caffeine-containing products targeting cellulite.

5.9 CONCLUSION

It is hoped that we have learnt from this chapter why skin is so important, how it works and why it deserves to be looked after and treated well. The complexity of skin goes far beyond the scope of this book; for more information on the science of skin, see *The Remarkable Life of the Skin: An Intimate Journey Across Our Surface*.[5]

A good-quality canvas is an important first step to any great painting. Cosmetics, like paints, are always applied to a substrate and in this case the substrate is skin. Healthy-looking, smooth, even-toned skin is the start to making all cosmetic products perform to their best ability. Holding back the visible signs of ageing has always been the holy grail, and skincare routines can play an important part in this elusive challenge. However, effective results from skincare treatment will only be achieved with regular use and, in order to ensure that users comply with this, the products must be enjoyable to use. Understanding how to produce well-formulated skincare products that customers will return to for many years is therefore a fundamental pillar of cosmetic science.

REFERENCES

1. N. Labban, H. Al-Otaibi, A. Alayed, K. Alshankiti and M. A. Al-Enizy, *Saudi Dent. J.*, 2017, **29**(3), 102–110.

2. T. O'Lenick and T. O'Lenick, PEG/PPG dimethicone structure and function, *Personal Care Magazine* April 2013, 1–5.
3. S. Barton, The Composition and Development of Moisturizers, in *Treatment of Dry Skin Syndrome*, ed. M. Lodén and H. I. Maibach, Springer, Berlin, Heidelberg, 2012, pp. 313–339.
4. Maison G. De Navarre, *The Chemistry and Manufacture of Cosmetics*, D. Van Nostrand Company, 1941.
5. M. Lyman, *The Remarkable Life of the Skin: An Intimate Journey Across Our Surface*, Penguin Books Limited, 2020.

CHAPTER 6

More Than a Smudge of Colour – The Science Behind Colour Cosmetics

CLAIRE SUMMERS*[a] AND PAULINE DUBOIS[b]

[a] Azelis UK Ltd, Hertford, UK; [b] C&R Packers, New Zealand
*Email: claire.summers@azelis.co.uk; pauline-dubois@live.fr

6.1 WHY DOES SOMETHING APPEAR COLOURED?

There are many factors that affect how something appears coloured. First, although it sounds obvious, there needs to be light. One then needs to have a colourant and, finally, to establish the way in which the surface of the coloured object reacts to the light that touches it. The process and variables are surprisingly complex, and it is hoped that through the following discussions we can 'shed some light' (no pun intended!) on the reasons why something appears coloured and what the implications are for all the various coloured cosmetic products you see in your home and on the shelves.

An object such as a lipstick appears coloured if it either absorbs or reflects a certain part of the visible electromagnetic spectrum (or light), as described later. These properties of

---

Discovering Cosmetic Science
Edited by Stephen Barton, Allan Eastham, Amanda Isom,
Denise McLaverty and Yi Ling Soong
© The Royal Society of Chemistry 2021
Published by the Royal Society of Chemistry, www.rsc.org

absorption and reflection are what cosmetic scientists play with to create the coloured products you may use every day. However, it really is all in the eye of the beholder.

Have you ever been in a nightclub when they switch the lights on at the end of the night? Everyone looks slightly less perfect when the darkness changes to 'glorious technicolour'.

The retina of the human eye is covered with around 130 million photoreceptors, 125 million of which have a rod shape and are called 'rods' and 5 million have a cone shape and are called 'cones'. When a light signal reaches the photoreceptors (rods and cones) (Figure 6.1), an electric impulse is generated. This impulse is then transmitted to the brain *via* the optical nerve. Rods are very sensitive, adapting to very low light levels, whereas the cones, which detect colour, are less adaptable. As a result, in the absence of light, colours become either muted or shades of white, black or somewhere in between.

Another characteristic when light is present is the way in which objects appear coloured by the way they interact with the

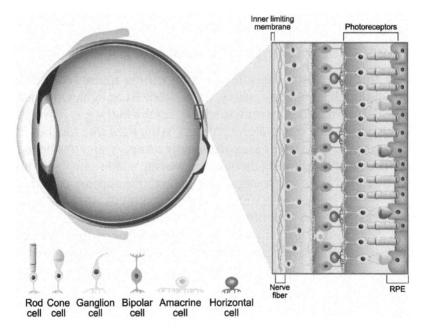

Figure 6.1 How the eye perceives colour. Cross-section of the human eye showing the retinal rods and cones that act as light sensors and transmit this information to the brain (© Shutterstock).

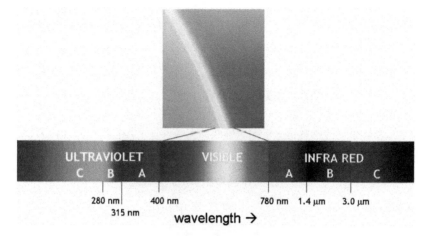

Figure 6.2 Colour and the electromagnetic spectrum. Light is part of the electromagnetic spectrum. This diagram shows the familiar rainbow of colours and shows the parts we do not see – ultraviolet (UV) and infrared (IR) light. These have some interesting properties in bringing about fluorescence but, apart from UV protection properties discussed in Chapters 7 and 8, are not discussed further here (Copyright © European Union, 2012; used with permission).

visible electromagnetic spectrum. The section of the electromagnetic spectrum or visible light that we can see is between 400 and 700 nm, which is a very small part (Figure 6.2).

If an object reflects all of the wavelengths from the visible electromagnetic spectrum, then it will appear white; conversely, when the object absorbs all of the wavelengths, then it appears black. In simplistic terms, we will see only the colour reflected and this is what gives an object its colour. The same light illuminates the apple and its leaf below (Figure 6.3), but because they absorb or reflect different wavelengths of the visible electromagnetic spectrum (rainbow) the apple and leaf have totally different colours. The leaf reflects the green part of the visible electromagnetic spectrum and absorbs all of the other colours from the 'rainbow' and the fruit reflects the red and absorbs all of the other colours from the 'rainbow'.

However, the next area we need to be aware of is the nature of the surface, and how it alters the colour seen. This is important as the eye can only see the object and colour of that object by the light that it reflects. There are two types of reflection, specular and diffuse reflection.

Figure 6.3 Colour absorption and reflection. As described in the text, absorption and reflection help us interpret the colour and texture of objects. The leaf absorbs all wavelengths except green, which is reflected back as the colour we know. The surface texture is a result of light being reflected in many different directions. The fruit absorbs most wavelengths of light except red and some other colours, which reflects back the apple colour. The fruit surface is smoother than the leaf, having a more shiny effect (© Shutterstock).

Specular reflection is the light you see from a mirror, a perfect reflection of what is in the mirror. In the case of the apple, this why we see a white sheen where the light is reflected back most directly. In cosmetics this is used to enable us to see the effect of pearlescent pigments. We shall return to this in more detail in Section 6.2.3 and Figure 6.10.

Diffuse reflection – where light is scattered back in many directions – allows us to interpret texture, such as the veins in a leaf. Diffuse reflection from pigments and powders is used by formulators to disguise or hide imperfections and wrinkles by changing the directions in which light is reflected back from the surface, a very useful and well-used trick of the cosmetic industry.

The quality and type of the light are also important, and vary as a result of many factors, not least of which are the time of day, time of year and geography. If the weather is always blue skies with no clouds to remove any colours from the visible electromagnetic spectrum, the light is known as north light as it would consist of the full visible electromagnetic spectrum, red, orange, yellow, green, blue, indigo and violet, but this is rarely the case, unless you live in warmer climates. To allow colour cosmetics scientists to judge colour objectively and repeatably, colour is viewed inside light boxes (cupboards) with specific light qualities to avoid colours being seen differently depending on the quality of the light.

When describing colour, everyone will have a different interpretation of a particular colour. For example, what would you call the colour of the shaded area shown in Scheme 6.1? Now try again with this shaded box as shown in Scheme 6.2 – would you use the same description? The box fill colour is identical in both cases, but changing the surrounding colour and text colour may influence your perception and description.

Another point to highlight here is that colour perception in people varies dramatically – sometimes termed colour blindness. Did you know that colour blindness affects approximately 1 in 12 men and 1 in 200 women in the world? Colour blindness is mainly due to an inherited genetic condition but can also be caused by chronic illnesses, accidents, medications or ageing. There are several types of colour blindness. 'Trichromats' have full colour perception, which means they see all colours in the visible electromagnetic spectrum. 'Anomalous trichromats' are

Scheme 6.1

Scheme 6.2

unable to see various wavelengths of the visible electromagnetic spectrum and finally 'monochromats' cannot see colour, so see the world through various shades of grey. Monochromats are common in the animal kingdom but rarer in humans.

For these reasons, 'measuring' colour *subjectively* using the human eye can be unreliable and it is preferable to do this *objectively* using some form of colorimeter. Their functioning is to give a colour a value objectively to allow it to be reproduced *via* an instrumental method (see Box 6.1).

BOX 6.1   HUE AND CHROMA

Colour scientists understand fundamental parts of colour in terms of 'hue' and 'chroma'.

Colour hue

This refers to the different elements of the spectrum and how they can combine. It comprises:

- five principle colours – yellow, red, purple, blue and green;
- five intermediate colours – orange, red–purple, purple–blue, blue–green and green–yellow.

These colours are illustrated in Figure 6.4.

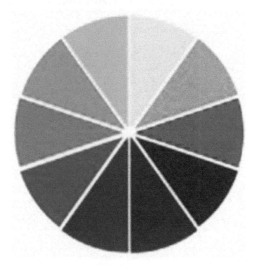

Figure 6.4   Colour hue.

Colour chroma

This indicates the *strength* of the colour and is measured on a scale of 0 to 16. One method of measuring the chroma and hue objectively is using the $L^*a^*b^*$ system, where L^* gives information about the black–white axis, which indicates the intensity of the shade of colour: the higher is L^*, the more intense the is colour. The a^* value gives the amount of red or green present in the colour. Finally, the b^* value gives the level of yellow or blue in the colour being measured. This allows values that are basically coordinates in the colour space $L^*a^*b^*$ which relates to a specific colour. This concept is visualized in Figure 6.5.

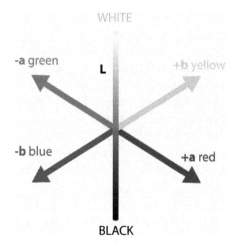

Figure 6.5 $L^*a^*b^*$ colour space for determining 'chroma' (© Shutterstock).

6.2 HOW CAN WE CREATE COLOURED PRODUCTS?

People often ask whether colours are chemicals. The quick and easy answer is 'yes' – everything around us is made up of chemicals and colours are no different. Colours for labelling purposes on ingredient lists are always referred to by their colour index or CI number. For example, titanium dioxide, which is widely used in many cosmetic formulations, will always appear listed as CI 77891.

There are many types of colours available in the formulator's palette, which we use to achieve the desired colour.

There is more than one way to get colour: synthetic dyes, natural dyes, inorganic pigments and organic pigments.

6.2.1 Dyes

Synthetic dyes are classified as chemical compounds that are soluble in a medium or range of media such as water, oils and alcohols. There are many different dyes that are used to colour toiletry products such as foam baths, but these are never used in colour cosmetics. Within colour cosmetics there is one group or dyes that can be used, which are referred to as eosin dyes. These have a staining effect and so were used historically for the first developed 'long-lasting' lipsticks. Care is required when formulating using these dyes as they can lead to allergic reactions in some consumers when used even at low levels.

The use of natural dyes in cosmetic formulations is currently limited; the strength and intensity of the colours they produce tend to be much weaker than those with traditional organic and inorganic colourants. Natural colours also have limited heat, light and pH stability. Another aspect that needs to be taken into consideration is that certain natural colourants have an associated odour. The most commonly used natural dye is carmine, also known as cochineal or carminic acid. Unlike many others, this has a very stable, strong and vibrant pink shade and is also used in the food and pharmaceutical industries (where it can appear as E120). However, some consumers object to the fact that it is derived from an animal – the cochineal beetle. This, together with some concerns about the safety of carmine in food and drink, have reduced reliance on this colour.

Other natural colourants are being looked at, including chlorophylls, which have the possibility to be shades of greens through to blue. Shades of orange from β-carotene, turmeric and paprika and pink from beetroot are other possibilities for the future. Research is still ongoing into these natural colourants, but it will take many years to achieve the same attributes as in the widely used inorganic and organic pigments in today's cosmetic formulations.

> **BOX 6.2 WHICH COLOURANTS ARE USED IN MY HAND WASH AND SHOWER GEL PRODUCTS?**
>
> Shower gels use dyes rather than pigments, with the definition of a dye being 'a colour that is soluble in the medium'. In the case of shower gels, which are composed of water and a surfactant (or the foaming active – see Chapter 2), the dye is soluble in the water but is used at very low levels, often below 0.001%. Blends of different colours can be used to achieve the desired colour. Many people keep their products on the bathroom windowsill and some colours can fade in sunlight. Very small amounts of UV filters can be included in the product or the packaging to prevent this happening in a clear glass or plastic pack.

When you look at the colours used in toiletry products such as foam bath, shower gels, toothpastes and lotions, they are soluble in the formulation or medium (in these cases normally water) and are referred to as dyes not pigments (see Box 6.2), but we can go into that further later on.

6.2.2 Why Pigments Are Crucial for Colour Cosmetics

A pigment is defined as a coloured chemical compound that is *insoluble* in the medium in which it is used. Pigments are classified according to their chemical make-up – the most important classes are outlined here.

6.2.2.1 Organic Pigments. These are composed of carbon, hydrogen, oxygen and nitrogen. They have bright, clean, vivid colours, as illustrated in the colour chart in Scheme 6.3, but some are, or can be, susceptible to interactions with other raw materials used in cosmetic formulations and the pH of the formulation can cause issues. These can be classified as azo compounds, which can be further split into insoluble azo, soluble azo, slightly soluble azo and soluble non-azo colours.

So, what is an azo colour? It is associated with compounds in which two adjacent nitrogen atoms are situated between carbon atoms (Figure 6.6), R–N=N–R′, where R and R′ can be either aryl or alkyl. The N=N group is called an azo group. The name azo

Scheme 6.3

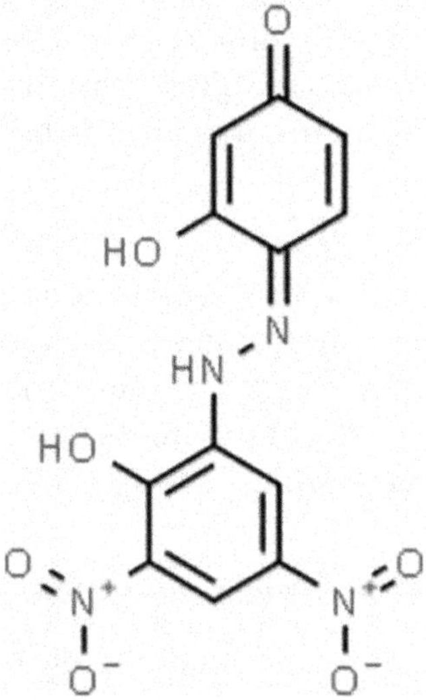

Figure 6.6 Structure of azo dyes. There are many variations on the basic structure. The N–NH link between the two six-carbon ring structures in this molecule gives the molecule its azo dye properties. The other atoms on the ring structure show that the dye can be charged. This allows many different types of atoms to attach, resulting in variations in the dye properties such as solubility.

comes from *azote*, the French word for nitrogen that is derived from the Greek *a* (not) + *zoe* (life).

There are three types of organic pigments: lakes, toners and true pigments.

6.2.2.2 Lakes. Various definitions exist to describe the word 'lake' but it is essentially 'an insoluble colourant produced by precipitating (attaching by chemical reaction) a soluble dye onto an insoluble substrate'. A lake is created when we take a soluble dye used in toiletry products and precipitate it onto alumina, *e.g.* Yellow 5 Al Lake (CI 19140). The precipitate that forms 'sticks' to the alumina substrate. This acts as a building block to create an insoluble pigment, making it suitable for use in colour cosmetics formulations. An issue that can be arise is that lakes are affected by extremes of pH when added to colour cosmetic formulations, which results in the soluble dye re-forming, a condition known as 'bleeding'. This 'bleeding' is the effect seen when people with lines around their lips (often as a result of ageing or smoking) apply lipstick that can then be seen travelling into these fine lines, giving the bleeding effect. This 'bleeding' is not seen with other pigments, so selection must take into account the target consumer. In addition to this problem, these types of organic pigment are relatively transparent and have poorer light stability than other organic pigments.

The two mostly commonly used lakes are Yellow 5 Al Lake (CI 19140) and Blue 1 Al Lake (CI 42090).

6.2.2.3 Toners. Toners are made in exactly the same way as lakes but instead of precipitating the soluble dyes onto an alumina substrate, other cosmetically approved metals are used to render the pigment more resistant to heat, light and pH. However, they remain susceptible to extremes of pH and can still result in different shades even when using the same toner.

A few commonly used toners include Red 6 Barium Lake and Red 7 Calcium Lake. The most interesting thing about these two colours is not just the fact one is yellow–red and the other blue–red, but that they are actually exactly the same colourant CI 15850. The resulting colours are totally different because the colourant has been precipitated onto different substrates (barium and calcium, respectively).

6.2.2.4 True Pigments. These are insoluble compounds that contain no metal ions and therefore it has not been necessary to precipitate them onto a substrate; they do not exist as dyes. True pigments are the most stable but unfortunately only a few

> **BOX 6.3 DECODING THE COLOURS**
>
> Identifying dyes and pigments can become complicated as they are used in so many different industries. In this chapter we have used the INCI (International Nomenclature of Cosmetic Ingredients) name together with the CI number to help you identify colours you may see on packs, for example, Black Iron Oxide (CI 77499).
>
> CI stands for Colour Index and CI numbers are used as identification numbers to differentiate colourants according to their chemical structures. CI numbers are referenced on a database managed by the UK Society of Dyers and Colourists and the American Association of Textile Chemists and Colorists. The Cosmetics Regulation (EC) No. 1223/2009, which applies to products sold in Europe, requires colourants to be listed with their CI number. Even though hair dyes fall into the colourant category when looking at their physicochemical properties, they are not listed in the Colour Index and will therefore be referenced by their INCI name.
>
> As you may have noticed, the names of the lakes and toners give away the identity of the metal used as a substrate – aluminium (Al); barium (Ba) and calcium (Ca). They also show which soluble dye has been precipitated onto these metals – Yellow 5, Blue 1, *etc.*
>
> You may also see the letters F, D and C used in some places and references. This coding has been used to show you that these colours have been approved for use in the different product categories Food, Drug and Cosmetic, respectively, under legislation in the USA.

exist. Red 30 (CI 73360) and Red 36 (CI 12085) are good examples of true pigments.

The scheme for identifying colours is explained in Box 6.3.

6.2.2.5 Inorganic Pigments. These are widely used in all colour cosmetic formulations and are composed of the elements carbon, hydrogen, oxygen and nitrogen just as those found in organic pigments, but they have additional elements including iron, titanium, calcium and barium, to name just a few.

More Than a Smudge of Colour – The Science Behind Colour Cosmetics 167

Scheme 6.4

Inorganic pigments are much more 'natural looking' or earthy colours – as illustrated in the colour chart in Scheme 6.4 – and they are normally quite opaque. They are much more compatible with other cosmetic raw materials, but a few have stability issues when used in acidic or alkaline conditions. When we talk about pH, it is important to understand that the pH scale goes from 0 to 14, with pH 7 being neutral, pH 0 being very acidic and pH 14 being very alkaline.

The most commonly used inorganic pigments are iron oxides. These consist of yellow (CI 77492), red (CI 77491) and black iron oxide (CI 77499). Brown iron oxides are sold but are just a blend of the yellow, red and black iron oxides. The pigments are very stable in all formulations and are easy to distribute within a formulation, are chemically inert (do not react with other chemicals), are not sensitive to solvents and are stable over a wide pH range from 4 to 10. Red iron oxide is actually obtained by the controlled heating or calcining (a heating, oxidizing or purification process) of yellow iron oxide. However, it is not just formulation stability that is important when choosing iron oxides. Black iron oxide is magnetic, which can cause issues in production if the manufacturing vessels are not earthed, making it difficult to clean vessels, and also some of the black iron oxide will be attracted to the metal manufacturing vessel.

The inorganic pigment chromium oxide (CI 77288) can range from a dull olive colour to a blue–green colour and has excellent heat and light stability.

Ferric ammonium ferrocyanide (CI 77510), also known as Iron Blue, is a very intense, opaque dark-blue pigment and is physically very hard, making it difficult to disperse; we will go into this later (Section 6.3). The formulator, however, needs to be aware that it is unstable at high pH, which results in a loss of colour intensity that is not reversible.

The ultramarines (CI 77007) have probably the most diverse colour range of all the inorganic pigments as they can range from blue to violet, pink or even green. The most commonly used type is Ultramarine Blue, but with all of the ultramarine range care must be taken always to keep the formulation above pH 7. If not kept above this pH, hydrogen sulfide will be produced, giving the characteristic odour of rotten eggs.

The final inorganic pigment worth a mention for needing to be treated correctly is Manganese Violet (CI 77742), which has a bright and clean purple or violet colour. This pigment is unstable at highly alkaline pH and will either turn black or disappear completely, and again this is not reversible.

6.2.3 Are Inorganic and Organic Pigments the Only Materials Used to Create Colour?

No, a number of other types of material are widely used in decorative formulations to enable a variety of colours to be achieved. Important examples are the pearlescent pigments.

6.2.3.1 How Does the 'Pearl Effect' Arise? It is probably well known that pearls, from oysters and other shellfish, develop over time as a gradual build-up of layer upon layer of something over a foreign body within the creature – usually a parasite, not the proverbial 'grain of sand'. The pearlescent effect is created when light is reflected in a preferred direction through these different layers, giving an effect that can have many names, including; gloss, lustre, glitter, glimmer and sparkle, to name just a few. All these names are normally dependent on the particle size of the pearl – which in the case of cosmetic pearlizing agents can vary from very small to very large.

However, this effect is not unique to shellfish. Historically, pearlescent pigments were based on guanine from fish scales, predominately derived from herrings and sardines, but the yield from the fish scales is only approximately 1% so for every 100 kg of fish scales you end up with 1 kg of pearl.

The first synthetically manufactured pearls were based on mercuric chloride and lead arsenate, which have known toxicity issues. In the 1930s, alkaline lead carbonate pearls were developed and widely used for the product of artificial pearls but were not suitable for use in cosmetic formulations.

The first synthetically produced pearls for cosmetics were created in the 1960s and were based on bismuth oxychloride and crystals were grown. The greatest limitation with this type of pearl is that they have a high lustre when wet but a low lustre when dry.

These pearlescent pigments consist of a substrate (building block), which has a colour attached to the surface to achieve a variety of effects.

Subsequently mica was discovered, a new substrate that was chemically and thermally stable. As it is used in formulations that are heated above 100 °C and mixed with other raw materials (or chemicals), it has good transparency and so could be used as a substrate for the creation of a new generation of pearlescent effects.

Mica, when ground, creates five fractions of different sizes ranging from very small (5 nm) to very large (200 nm), so is very flexible for the creation of different sized pearlescent pigments to give different effects from lustre through to sparkle.

Mica led to the first developed cosmetic pearl pigment for the cosmetic formulator, referred to as titanated mica pearls. They are white pearls with no coloured pigment present and are based on the coating of the mica *in situ* with titanium dioxide (as in Figure 6.7).

The next step in the development of these mica-based pearls was the creation of interference pigments that, although they contain only titanium dioxide, appear coloured when added in a product or on the skin.

Before explaining what interference is, you need to understand how interference is created. This is something you will be familiar with if you have seen diesel fuel or oil floating on a puddle of water at the service station. Interference is the ability to use light to create colour. The light waves reflected back from the top and bottom surfaces of the oil (or other material) 'interfere' with each other, causing certain wavelengths to be cancelled out (absorbed), leaving other parts of the visible electromagnetic spectrum to be reflected back. Varying the thickness of the material changes the colours. For cosmetic interference colours, varying the thickness of the titanium dioxide on the mica substrate allows colours to be created, as shown in Figure 6.8.

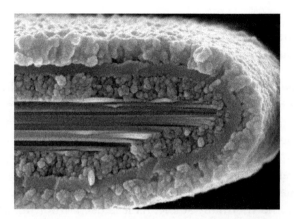

Figure 6.7 Mica coated with titanium dioxide. The high magnifying power of a scanning electron microscope shows the fine detail of how a coating is layered onto a substrate to create a pearlizing effect. The central plates are the mica and the surrounding crystalline structure the titanium dioxide (Used with permission, © Merck).

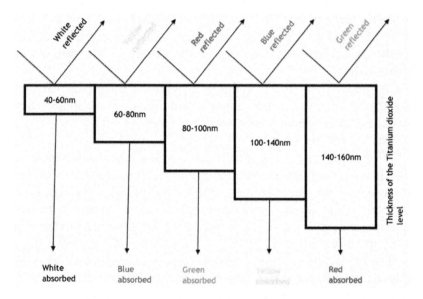

Figure 6.8 Interference effects, illustrating how light bouncing back from the top and bottom of a microscopic layer allows these reflecting rays to 'interfere' with each other's wavelengths to produce different parts of the spectrum.

> **BOX 6.4 WHAT IS THAT LUSTRE IN MY SHAMPOO?**
>
> The lustre or pearly effect seen in some shower gels or shampoos can be achieved using pearlizing agents. Unlike some of the pearls used for skincare, many of these are the result of organic fatty materials such as glycol distearate (also known as ethylene glycol distearate, EGDS) and glycol monostearate (also known as ethylene glycol monostearate, EGMS), produced by esterification of stearic acid and ethylene glycol. These materials crystallize during the cooling phase of the manufacturing process and the suspended crystals create an optical *interference* effect, giving a lustre effect to the product.
>
> Pearlescent pigments can also be used to create a shimmery effect, but care needs to be taken in selecting the correct particle size to achieve the desired effect and also using a stabilizer that gives enough suspension for the selected pearlescent pigment.

These interference pearls are often used by cosmetic formulators to disguise skin imperfections. Adding interference green on red skin is one example, causing the red imperfection to appear reduced as we are using light to hide the redness.

You will also be aware that this 'pearlizing effect' is seen in other products, such as shampoos, but the source of pearlizer is different (see Box 6.4).

6.2.3.2 Creating Colour Using Light. This ability to use one wavelength of light to cancel out another colour allows us to use additive colour mixing. From Figure 6.9, you can see that by mixing pure colours of light – red, green and blue – you can obtain white light, a method that is commonly used in the theatre world.

The next step in developing mica pearlescents was to create coloured pearls by coating the mica substrate with other pigments such as yellow iron oxide to create gold pearls, or other iron oxides to create metallic effects. These were then combined with the technology of interference to create coloured pearls with an interference effect.

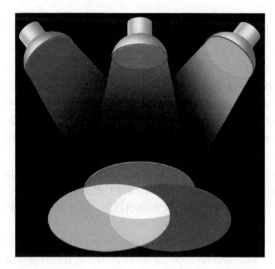

Figure 6.9 Additive colour mixing. Mixing primary colours – red, green and blue – can produce both white and the secondary colours – cyan, magenta and yellow. This technique is also used in stage lighting in the theatre (© Shutterstock).

Other synthetic substrates are now available that are cleaner and more translucent and therefore with even more of a pearlescent effect. These include synthetic fluorphlogopite and calcium aluminium borosilicate.

6.2.3.3 Light-diffusing or Soft-focus Pigments. Another variety of pigment that is commonly used in colour cosmetics is non-pearlescent fillers. These raw materials are created from a substrate with a pigment coating but have no 'pearl' effect. These materials act to change the perception of colour *and* the texture of the surface to which they are applied, creating a soft-focus effect. As can be seen from Figure 6.10, rather than bouncing the light back at a perfect 'mirror' angle, they cause the light to be scattered in various directions. Since the skin surface is highly textured, natural light will bounce back in different directions from the different features – lines and wrinkles, raised follicles and spots, for example. By increasing the number of directions in which light is reflected back, these soft-focus pigments help to mask any such skin imperfections.

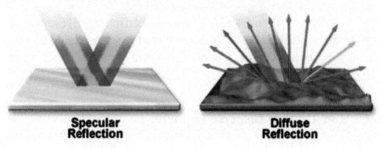

Figure 6.10 Specular and diffuse reflection. When light reflects from a totally smooth surface, all the rays are reflected in the same direction – the closest we come to this specular reflection in cosmetics are a nail varnish surface and lip gloss. In most other cases, the light hits a surface that is more or less rough – the skin itself is finely textured. In this case, 'diffuse' reflection occurs because the light is scattered in many different directions (Used with permission, Copyright © Olympus).

6.3 ARE PIGMENTS EASY TO USE?

In one word, *no*! To achieve a uniform and reproducible colour in the product and on the skin, pigments need to be evenly distributed in the product. Understanding the properties of the different types of pigment and the best way to incorporate and use them successfully is an essential skill for any formulator.

Often the 'pigment', irrespective of the type being used, is the most expensive part of any decorative cosmetics formulation and critical in achieving the desired effect. Therefore, it is essential that the pigment is as fully dispersed (evenly distributed) as possible. Each pigment type requires different equipment, so it is worth going into the dispersion techniques. It is also necessary to know what to avoid as poor dispersion can lead to instability of the formulation and poor consumer experience.

Decorative cosmetics are unique within personal care formulating. They involve the widest range of different formulation bases of any personal care category. For example, in lip products, bases can vary from wax systems to gelled oils, powders and water-based jellies, to name just a few. Foundations can be emulsions – oil-in-water (o/w), water-in-silicone (w/si) and even water-in-oil (w/o) – anhydrous wax systems (water free) or even powders. Eyeshadows and blushers can take the form of powders, creams, soufflés, waxes and 'cream-to-powder' formulations.

Each of these formulations has specific requirements and considerations when it comes to creating the desired colour effect and, as we shall learn later, the desired textural and sensorial properties.

6.3.1 How Do You Disperse Pigments?

Pigments need to be dispersed if they are going to provide an even layer of colour on the skin. This dispersion depends on the type of pigment, so we can first look at inorganic and organic pigments and then the pearlescent pigments.

Inorganic and organic pigments are supplied in powder form, so the greatest challenge is the dispersion of an insoluble dry pigment into a range of formulation types. These can be liquids, such as emulsions if making a foundation, solids such as waxes when creating a lipstick, or powders such as an eyeshadow, and for each you need different dispersion techniques.

An essential consequence of not properly dispersing these pigments is that it can result in loss of stability or functionality, which will result in a decrease in customer confidence. From a colour point of view, if the pigment is not dispersed then more is needed as the colour will not have the same strength, which impacts further on cost.

In an ideal world, the dry powder inorganic and organic pigments would be in the form of primary (individual) particles so they would immediately be dispersed; unfortunately, this is not the case.

Primary particles would be individual pigment particles, but normally the powder will be a blend of flocculates, agglomerates and aggregates (Figure 6.11) that need to be made smaller – ideally as primary particles, but realistically aggregates.

One aspect to highlight here is that after dispersion has occurred it is very important to stabilize the dispersion since the dispersing process is reversible and the pigments will return to their undispersed state if left without structure. For example, they might sink to the bottom. A variety of mechanisms are used to stabilize the dispersion, ranging from thickeners to increase the viscosity of the whole pigment blend to dispersing aids that surround the pigments to stop them joining together.

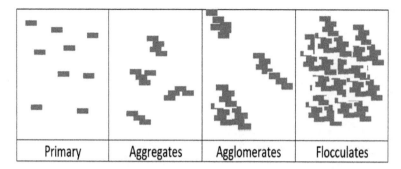

Figure 6.11 Particle structures. As described in the text, 'primary' particles rarely occur; more often they exist as flocculates, agglomerates and aggregates.

Figure 6.12 Creating coloured emulsions. The homogenizer breaks down large agglomerates and emulsion droplets into smaller ones using the vortex from the central propeller (as shown by the arrow). This recirculates them through the fine mesh (Used with permission Copyright © Silverson Machines UK Ltd).

For both liquid and wax systems, a high-shear mixer (very high speed combined with pressure) such as a Silverson is an essential piece of equipment. The vortex created by the head of the high-shear mixer breaks down and disperses the agglomerated particles (Figure 6.12).

Figure 6.13 Triple-roller mill. Milling the pigments presses them apart, as shown by the size of the red dots in this schematic diagram.

Incorporating organic and inorganic pigments into an ester or castor oil for a lipstick formulation requires a triple-roller mill (Figure 6.13). The three wheels turn in different directions and the pigments are essentially pressed apart. The greatest drawback of using a triple-roller mill for dispersing the pigments is that product losses can vary between 3 and 35% of the batch depending on the colour and the flow of the dispersion.

For powder formulations, a grinder needs to be utilized to break the agglomerates down physically using blades. One piece of equipment that is used for this is a hammer mill, as shown in Figure 6.14.

Pearlescent pigments, however, do not follow any of the same rules as organic and inorganic pigments as they already exist as primary particles so dispersion is not necessary. It is more important to achieve a homogeneous (even distribution) mix in the final formulation and just low-shear mixing is required with a propeller stirrer in a liquid or anhydrous wax formulation to achieve this. For powder formulations a ribbon blender is used, which simply turns the powder to mix the raw materials together.

However, the most critical aspect when adding pearlescent pigments to any formulation in production is to avoid high shear, which has the potential to destroy the pearl. The larger the pearl, the greater is the risk, so the use of high-shear mixing of

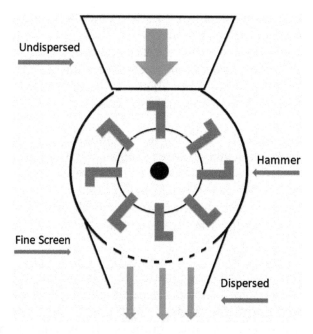

Figure 6.14 Hammer mill. A large-scale grinder used to break down agglomerates for a powder formulation. This has a similar action similar to a coffee grinder that you might use in your kitchen – but on a much larger scale.

liquids and hammer mills for powder formulations is never advised in these situations.

6.4 WHY THE TEXTURE OF COSMETIC FORMULATIONS IS SO IMPORTANT

As with all cosmetic products, their sensorial properties are as important as the effect they create. Although no-one will buy a foundation that is the wrong colour for them just because it feels amazing, they may reject the right colour on the basis of its poor skin feel. A brilliant coloured nail varnish will not find favour if it takes hours to dry or remains sticky all night! We can create a matte lipstick that has no gloss, but it will feel horrendously dry, or a lipstick with long-lasting properties but which has a sticky, unacceptable texture when applied.

To address *all* of the consumer's needs, a formulation requires a balance of raw materials that are designed to work together to satisfy all the sensorial and performance requirements demanded of the different product types.

Due to the variety of cosmetic formulations, there is a wide range of raw materials available to the formulator; some of these have been mentioned in other chapters – particularly Chapter 5, which looked at skincare formulations:

- *Film formers:* These are often polymers, such as trimethylsiloxysilicate, or waxes such as *Oryza sativa* (rice) bran wax, that aid in the adhesion to the skin, so improving the wear properties of the final formulation.
- *Structurants:* Soft beeswax (cera alba) and hard waxes (microcrystalline wax) for structure and stability in a wide variety of formulations.
- *Emulsifiers:* These are used where water and oil or water and silicones need bringing together for a formulation. They can be used to help disperse pigments and to help deliver the end user's skincare needs.
- *Thickeners:* For example, gums such as guar gum, and polymers, such as carbomer, are used to aid stability, modify the texture and change the application characteristics of the formulation. See Box 6.5 for more on thickeners.
- *Binders:* For example, pentaerythrityl tetraisostearate (PTIS). These allow the compression of powder formulations but also affect how well the final product will adhere to the skin.
- *Others:* Additionally, a wide range of esters, oils, humectants and emollients are used to modify product texture and feel, application characteristics and stability.

Let's look into each type of formulation and discover what raw materials we would generally expect to the used, and why.

6.4.1 Foundations – Are They More Than Just Colour?

First we need to determine what the requirements are for a foundation. These vary depending on the needs of the end user, but first and foremost the colour and consistency of it are vital, followed by texture, application characteristics (including spreadability), drying time and wear attributes.

> **BOX 6.5 YOUR 'BFF' – 'BEST FOUNDATION FRIEND' OR 'BENTONE FOUNDATION FRIEND'?**
>
> The greatest challenge when formulating any foundation is stabilizing the pigments to prevent sedimentation. A popular choice that can prevent the sedimentation is the incorporation of colloidal additives (or stabilizers). These stabilizers will create a thixotropic system – 'thick' gels which become 'thinner' if you stir or shake, then later return to a thick system. Foundations will be strong enough to resist pigment gravity but thin down to allow the product to be pumped or squeezed from a tube but then not drip when applied to the skin. Such a physical network within the formulation is achieved by using organoclays. Organoclays are common clays (bentonite, montmorillonite, *etc.*) comprising silica-based inorganics and charged organic compounds. Their ability to form microscopic 'house-of-cards' structures to hold things together and break and re-form like quicksand is what makes these materials special.

The natural colour of skin depends on several factors, for example the skin thickness, location of blood vessels and capillaries and the amount of melanin it contains. It is also influenced by surface texture. Diffuse and specular reflection from the surface, and also the quality of light absorbed and reflected back from deeper layers of skin, all contribute to our perception of skin colour.

Foundation products are by far the most popular colour cosmetic product for the face. They are used to achieve various effects, including 'mattifying', 'radiance enhancing' and giving the skin an even tone. Foundations can be a variety of emulsion types: w/si emulsions are used for long-lasting formulations, o/w emulsions are for more basic general foundations and w/o formulations are generally used for extremely dry skin types where moisturization is needed. They will all contain an emulsifier or emulsifier blend, suitable for the emulsion type and also the requirements of the other raw materials in the formulation and any skincare considerations. A foundation is, in simplistic terms, a suspension of pigments that are matched to the required skin tone, so the stability or rheological yield value of a foundation is key to a stable formulation. Yield is needed in a formulation as it gives an internal structure that traps the pigments throughout the formulation. Without this yield, which raw materials such as xanthan gum in an o/w emulsion will give to the formulation, the pigments will either sediment (drop to the bottom of the formulation) or undergo a combination of coalescence (grouping together) and creaming (floating to the top), and neither of these instabilities is acceptable for the consumer or formulating chemist.

The colourants used are the inorganic colours, primarily the iron oxides, but if skin correction is needed then other inorganic colours are available and, if desired, pearlescent pigments to achieve a specific effect. With these it is possible to create the variety of skin tones suitable for the wide diversity of skin colours expressed in different ethnicities. One aspect to highlight here is the use of the 'white' pigment. For white skin tones titanium dioxide is used, but this needs to be switched to zinc oxide to avoid creating the appearance of 'ashiness' on darker skin tones.

Other common raw materials also present in foundations include pigment wetting agents, fragrances and preservatives. Wetting agents are a type of emulsifier/surfactant (described in Chapters 2 and 5) that aid the dispersion of the inorganic pigments. The role of fragrances and preservatives is covered elsewhere (Chapters 7 and 8, respectively).

Foundations do not just take the form of emulsions, and anhydrous (without water) wax-base formulations are also available, which are generally a mix of powders such as talc, starch or clay, with iron oxides in a wax and oil base. These can be referred to as cake foundations, cream-to-powder or simply a hot pour foundation. The key raw material in this formulation is the balance of powders, predominately starch, which gives a product's unique powder-like skin feel.

Another product often associated with foundations, due to their function, is concealers, which are formulated using the same raw materials but with much higher pigment levels. For a foundation, generally between 5 and 10% of inorganic pigments is used, but for a concealer it is normally in the 35–40% range. Concealers are generally wax based, as stabilizing that level of pigment in an emulsion to avoid sedimentation, coalescence or creaming is a challenge.

> **BOX 6.6 FESTIVAL READY – TIME TO SHINE**
>
> The microscopic particle size of a pearlescent pigment determines the effects on the skin – the larger they are the more they will sparkle. For festivals, the largest of pearls (irrespective of type) are selected to give the 'wow factor' so often desired. Historically, polyethylene glitters were used but these are microplastics, so are no longer used in either cosmetic (*e.g.* eyeshadows) or toiletry products (foam baths). They also have poor consumer acceptance, and Bioglitter® is becoming valued as a suitable alternative.

Tinted moisturizers or creams with hidden colour are also worth adding at this stage as they will have a very low level of pigments, or sometimes encapsulated pigments are used so that the cream is white when applied to the skin but colour then appears as it is applied and rubbed on, breaking the encapsulating layer and revealing the colour. This is also an area where additive colour mixing is common – yellow or green colours are often used to counteract skin redness or areas of uneven pigmentation.

The final aspect of the formulation that must be discussed is inclusion of other ingredients and functions to create a unique selling point (USP). Common approaches are the inclusion of UV filters to provide sun protection, adding anti-ageing or anti-redness actives, discussed in Chapter 8, through to the use of pearls to enhance the perception of skin radiance.

Some products also include glitter materials for more dramatic effects (see Box 6.6).

6.4.2 Powders – Simple Yet Surprisingly Complex

First we will look at the attributes that are required for a powder cosmetic formulation. The first thing to consider is the colour, using the correct effect from the selected pigments, both inorganic and organic pigments dispersed fully throughout the formulation. Questions that a formulator will be asking themselves include the following:

- Are the pearlescent pigments used giving the desired effect?
- Is the product smooth and easy to apply?

- Does the product work well with the intended applicator?
- Does the product adhere well to the skin when applied?
- Is the coverage as intended?

The attributes will also depend on whether the product is pressed (into a compact or godet) or is loose. All these considerations will determine the correct selection of raw materials in the formulation.

The key raw material in most powder formulations is *talc* – chemically hydrated magnesium silicate. This use is important in colour cosmetics because talc has several characteristics that make it ideal for this formulation type. First, it is white to off-white and it has good slip – the ability to reduce the resistance between two surfaces, in this case skin and skin. This is something that would be important where an eyeshadow is applied to the eyelid using a finger, for example. Talc also has low coverage, which allows it to look more natural than something that is opaque, such as titanium dioxide. In addition, it has similar shine properties to those of the skin, which adds to the natural appearance. The particle size of the talc used is critical, as too small or too large and the finger will perceive the talc to be gritty owing to our sensorial perception; 6–8 μm is the perfect particle size.

Starches and modified starches are now being used in combination with or as alternatives to talc; an added attribute is that normally starches are very efficient at absorbing oil on the skin.

Other chemical materials that can be used to modify skin feel include silica, nylon, mica and bismuth oxychloride, to name just a few.

Although they are essentially free from water, all powders need to withstand microbial challenge. Adding preservatives can ensure that the product is protected from what we transfer from our skin to the product when using it, in addition to preventing microorganisms present in natural materials such as talc from becoming harmful to health. See Section 9.2 in Chapter 9 for more details on this.

Binders are the key to both pressed and loose powder formulating. For loose powders, binders help the product to adhere to the skin and not immediately fall off. For pressed powders, the binder is more critical as the stability of the pressed powder is critical in the functionality of the formulation. Binders can be singular or a blend of several, and they can be oily liquids (capric/caprylic glycerides) or powders (magnesium stearate), or combinations of both. With pressed powders, the binder level needs to be adjusted when other raw materials are incorporated into the formulation. The most critical issue arises when high levels of pearlescent pigments are used; these prevent the binder from working as efficiently. An incorrect binder level can lead to flaking of the formulation, with the formulation being too hard (in its packaging), and glazing (becoming shiny and no powder coming off), with the formulation being too soft or the powder breaking if dropped.

6.4.3 Lipsticks: More Than Lip Service

What we need from a lip product is for it to be dermatologically safe – the raw materials need to be approved for potential ingestion, in addition to having an acceptable taste and odour. The required application characteristics for all lip products are that they must be smooth and non-greasy with consistent colour and use characteristics. They should have a varying degree of wear characteristics depending on the claims of the formulation and the type of lip product. If the lip product is a stick, then the structure of the stick or bullet is critical to ensure not only that it withstands the pressure from application onto the lips, but also that it maintains the stability of the waxes and oils within the crystalline structure. The wax blend is finely balanced to create a stable crystalline structure that prevents 'blooming' of raw materials within the formulation. Bloom is the result of a weakened structure that allows oils or waxes to 'escape' and come to the surface. A good example of a bloom is when chocolate is left out in hot weather and acquires a white cocoa butter bloom on the surface.

Formulations for lip products are a fine balance of waxes, oils and colourants and the combinations are infinite, which is why stick products are one of the most challenging of all colour cosmetic formulations.

The oils within lip formulations are more than just for feel and gloss characteristics and are an essential part of the formulation to aid in the suspension stability of the pigments within the formulation, either during production or post-production.

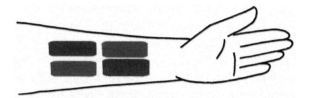

Figure 6.15 Colour matching. How test colours are applied to the forearm to assess colour match.

6.5 THE ART OF COLOUR MATCHING

Colour matching is both an art and a science, as you need to appreciate what all the pigments contribute in addition to how they work together.

The most common way to assess the colour match is to apply a product onto the skin, usually the inside of the arm, and compare it with the benchmark (Figure 6.15). A benchmark, or standard as is it is commonly called, is the colour that you want to match. If comparing with a similar product format or a previous colour match, then both shades should be applied to the skin in a cross-match position. This method can also be used for quality control purposes on batches produced in the factory.

When developing skin tone shades, it is important to assess the colour on a subject with the same skin tone as the intended consumer. Alternatively, synthetic skins can be employed; these replicate the skin texture and are available in various skin tones.

Another way to evaluate a shade match is by using a drawdown card. The product is spread uniformly on a card using a calibrated bar-type applicator or a Hegman gauge; these are available for many industries. This method is mainly used in the assessment of nail varnishes and allows the assessment of the coverage and shine of the applied film.

6.6 CURL UP AND DYE?

Now you know more about the chemistry of colour, this chapter would not be complete without considering hair colourants. Like hair styling, our human desire to enhance hair colour has a long and fascinating history. The use of extracts of henna (Figure 6.16), as evidenced from its presence in Egyptian tombs,

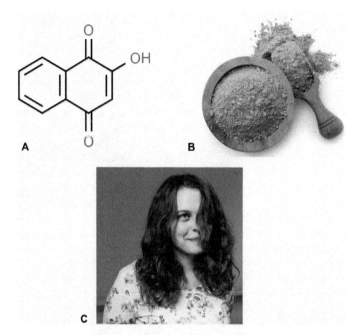

Figure 6.16 Chemistry and colour in hair dyes. The structure of lawsone from henna (A) shows the two rings of six carbon atoms with oxygen and hydroxyl substituents on one of them. (B) This chemical structure absorbs all wavelengths and reflects greenish wavelengths in powder form (© Shutterstock). (C) When it is prepared as a paste and reacts with hair, the absorption and reflection change slightly to create the brown colour. This also shows how new hair emerging from the root retains the person's natural melanin colouring (© Shutterstock).

woad (indigo dye), used by ancient Britons and Persians, and extracts from oak nut galls and walnuts in Asia, show that the chemistry of colour and adornment has been an important human interest for centuries. The chemical structures of the active materials from these natural sources – especially the brown colourants lawsone, pyrogallol and juglone – formed the basis of hair colourant development into the twentieth century (see Figures 6.16–6.18).

Although some dyes attach to the hair surface, they tend to be less intense and less effective, so getting a dye to colour the whole hair fibre is desirable. It will be clear from Chapter 3 that one of the greatest challenges here is getting dyes through the protective outer cuticle and into the hair. Since people often

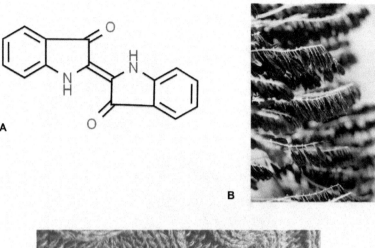

Figure 6.17 Chemistry and colour in hair dyes – indigo. (A) Indigo also has complex carbon ring structures but in this case two six- and five-carbon complexes are linked together. (B) Dyer's woad (*Isatis tinctoria*) seed pods hanging on branches. The colour produced by the indigo molecule can be extracted (© Duncan Smith/Science Photo Library). (C) This shows the similarities and differences between hair and textile dyeing. In wool dyed with indigo, the absence on melanin in sheep's wool allows the full colour to be displayed. The effects on skin and hair will be more muted as a result of the background melanin colour (© Shutterstock).

want a hair colour that is very different to their natural tones, ensuring that the intended colour is retained for an acceptable period is another challenge. Add to that the fact that many people have a blend of different coloured hair due to the mixture of brown eumelanin and red pheomelanin, and you begin to see another challenge, especially if you want to achieve a 'natural look' rather than a uniform block of dye.

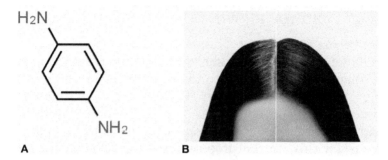

Figure 6.18 Chemistry and colour in hair dyes – *p*-phenylenediamine (PPD). (A) The PPD molecule shows some similarities to lawsone and indigo. The single six-carbon ring molecule is a smaller entity that may enter the exposed cuticle more easily. Once inside, the nitrogen atoms can form bonds with the hair to produce the classic dark browns. (B) This 'before and after' photograph illustrates a uniform colour across all hairs, including grey ones when using a PPD-based dye (© Shutterstock).

For greater detail of the chemistry used to create and the hair colour options, see https://www.compoundchem.com/2015/05/14/hair-dyes/.

The late nineteenth to early twentieth century saw an increased understanding of the chemistry of hair and wool dyeing (see Box 6.7). The role of oxidation is important. Not only does oxidation lighten the hair, it is also involved in the formation of some dyes that are often called permanent dyes. This understanding, together with a knowledge of other factors that could influence the dyeing process, particularly pH, all led to the development of a greater range of colours than the starting natural sources alone.

Most early hair dyes were applied by professionals – they could learn how to control the process and use their creativity to produce different effects. However, the desire for home-use products grew and this meant that ease of use was another factor behind the design of hair dyes.

6.6.1 Oxidation and pH

Hair colouring products fall into two general chemical categories – those based on oxidation and uncoloured dye precursors and those that use a ready-formed coloured 'direct' dye.

Reactive oxygen, in the form of hydrogen peroxide, for example, in combination with a high pH, which causes the hair to

> **BOX 6.7   DID YOU KNOW?**
>
> **Wool *Versus* Hair Colour**
> You'll remember from Chapter 3 that hair is very similar to fur and other mammal coats – so many advances in hair dyeing came from the textile industry. However, the one difference between dyeing wool and hair is that the wool is plunged into a bath of dye. Even the most vain people would be unlikely to consider this, so the proportions are reversed – a small amount of dye is put onto the hair!
>
> **Hair Colour Business**
> L'Oréal began when a young chemical engineer in Paris, called Eugène Schueller, started to make hair colouring products in his flat during the night in 1907. Schueller's first product was called Aureole (from the French word for halo, *auréole*) and he called his first company la Société Française des Teintures Inoffensives pour Cheveux (the French Harmless Hair Dye Company), which later was renamed L'Oréal.

swell and 'open up' the cuticle, causes a bleaching effect on melanin in the hair, leaving it more porous and susceptible to colouring. Although 'bleached blonde' alone may be the desired effect, lightening the hair and then applying colour is a good way of achieving extreme effects – this is a two-step colouring process that many readers may have experienced.

However, the oxidative mechanism can also be included in a one-step process. Lightening the hair *and* allowing uncoloured dye precursors to form a coloured pigment within the hair have some advantages (*e.g.* good grey coverage), in addition to convenience for home use. The uncoloured precursors form larger, coloured reaction products that become trapped in the hair.

The use of small hair dye precursors dominates the hair colouring market today. One reason is that, when carefully formulated with an oxidizing agent in the presence of ammonia at about pH 10.5, almost any hair colour can be achieved by careful selection of anything up to eight precursors. They also work extremely well at colouring the hair in a permanent fashion – variations in colour longevity can be achieved by varying the pH and oxidizing conditions and length of exposure.

The most important and best-known dye precursor is PPD (*p*-phenylenediamine), which is the basis of very effective permanent dark colours when used in combination with other precursors known as couplers (see Figure 6.18). PPD and a range of other precursors have been shown sometimes to cause very serious allergic reactions. In addition to formulation improvements to address this issue, including modifying the potentially aggressive formulation conditions, recommendations for testing a small area of skin before use are seen as useful measures to reduce the risk of an allergic reaction.

Some alternatives to ammonia, such as sodium carbonate or monoethanolamine, can give a less permanent effect and, together with less peroxide, give less background lightening. The result is a less noticeable effect as the dye washes out.

Unreactive coloured 'direct' dyes can give a 'semi-permanent' effect. This is where the colour tends to last for a few weeks or about five washes or, alternatively, allows temporary effects that wash out easily to be created. Which effect is achieved is determined by the chemical nature of the dye, its size and the nature of the formulation, for example the pH. Dyes that give a semi-permanent effect tend to be smaller and are used at higher pH, which facilitates penetration of the hair. Larger dyes that have a positive charge tend to stick to the surface of the negatively charged hair. Because the hair becomes more negative as it becomes damaged, for example as a result of repeated lightening, these dyes can cause very uneven wash out in different hair types. The use of direct dyes is much less predictable than that of oxidative dyes.

Even in the absence of allergy, it is well recognized that oxidizing high-pH conditions, together with the use of surfactants to deliver the system, can cause some hair damage. This helps explain the problem that many hair dye users have, namely that using a dye on already damaged hair will change the effectiveness of the process and affect the desired colour outcome. The use of protecting components within the hair colorant systems is a common way of reducing the hair damage that might be caused. They can also help maintain the colour intensity over time.

6.6.2 Natural Dyes

With all the issues discussed above, you may be asking why henna and other natural dyes, which are direct dyes, are not used more often. There are several reasons, not least of which is the lack of depth of colour. However, as with natural skin dyes mentioned in Section 6.2.1, it is also their instability and tendency to fade after UV exposure or due to regular washing that make them a less attractive choice for many people. Some improvements have been achieved by combining these natural dyes with metal salts, in a process known as mordanting. This is common in textile dyeing and involves metal ions attaching to the fibre and the dye, thereby binding them together. This helps reduce wash out.

Other ways of overcoming the deficiencies of natural colours have been to blend them in combination. However, here another issue arises – *safety*.

The safety of hair colourants is of the utmost importance; the combination of very reactive ingredients and the potential for some of them, including natural dyes, to produce an allergy (in addition to other toxic effects) have resulted in a carefully controlled list of permissible colours. Adding new natural colours to those permitted requires many years of research and safety assessment; the emergence of new colours into the market is not an everyday occurrence!

One consequence of all these factors is that the creation of hair colours is one of the most technically demanding and complicated types of formulation work. There is insufficient space in this book to go into all these complexities, but those interested will find sources of further reading at the end of the chapter.

6.6.3 Temporary Hair Colourants

One way to reduce potential damage to hair and achieve an interesting colour is to use one of the many 'temporary hair colourants'. These formulations have much more in common with the skin colourants described earlier in the chapter: spraying or painting onto the hair a film of pigment or dye. These have the advantage of being more easily removed than some of the other systems discussed above – however, getting caught in a rain shower may cause you to think differently about your choice!

6.7 CONCLUSION

Creating formulations that add colour to a person's cosmetic choices is both an art and a science. Understanding the chemistry of the colouring process provides the cosmetic scientist with a 'palette' to create a breadth of possible colours to mask or enhance their bodies. The factors that need to be considered to deliver the benefits and claims for any cosmetic formulation, whether it is a hair dye, a lipstick or a foundation, can by accounted for using the scientific principles within the base formulation. The balance that the formulator has created is the colour, which often is the greatest reason for purchasing a new cosmetic product, but the second purchase will depend on it performing well, and up to expectations. Understanding the way in which colour is used by a cosmetic scientist allows the development of a wide range of formulations with various different effects. This chapter has given you the basic understanding on how creating colour cosmetics is more than just a smudge of colour.

FURTHER READING

R. Christie, *Colour Chemistry*, Royal Society of Chemistry, 2nd edn, 2015.
L. Eldridge, *Face Paint: The Story of Makeup*, Abrams, 2015.
J. F. Corbett, *Hair Colorants – Chemistry and Toxicology*, Cosmetic Science Monographs No. 2, Micelle Press, 1998.

CHAPTER 7

Follow the Scent – The Science Behind the Fragrance in Products

V. DANIAU

Parfum Parfait Ltd, UK
Email: info@parfumparfait.co.uk

'You see, a fragrance arouses the mind'

From the poem *À Une Jeune Femme* by Victor Hugo

What the writer is telling us is that odours have a very powerful influence on our mind and behaviour, as many authors and experts have spoken about in their writings. Yet the mechanisms related to our sense of smell are still under research and much is left to discover.

In this chapter, we shall take a look at the mechanisms of olfaction, then follow with an overview of the creation of fragrances, in order to understand better how they enhance and impact our lives.

7.1 SOURCES AND MECHANISM OF ODOUR FORMATION

Of all our senses, smell is probably the one that remains the most mysterious. Olfaction mechanisms are still being actively researched by scientists and major discoveries have been made in this millennium. Researchers are exploring how precisely we perceive odorants, how the information is processed and how our brain makes sense of it all.

Let us begin by exploring the key mechanisms of our olfactory system and how it influences our behaviour.

7.1.1 Mechanisms

Smell is described as a 'chemical sense', because it is based on our perception of chemical variations in the environment we are in, more specifically in the variation of the concentrations of volatile molecules in the air that surrounds us. These volatile molecules can be detected in the nose, *via* olfactory sensory neurons.

Olfactory sensory neurons are nerve cells that communicate with other cells. They line the olfactory epithelium, which is a specialized tissue located at the back of the nasal passage. Tiny hairs called cilia are attached to these neurons; they contain receptors that recognize and bind to odour molecules. When we breathe, chemicals in the air are absorbed by the olfactory mucosa (mucus in the lining of the nasal cavity) and bind to those receptor sites on the cilia (Figure 7.1).

The cilia of these neurons protrude into the nasal mucosa (an odorant-binding receptor) that induces a signal in the cell and changes the membrane's electrical potential. This sends an electrical impulse, creating an odour 'message' that is then transmitted to the brain.

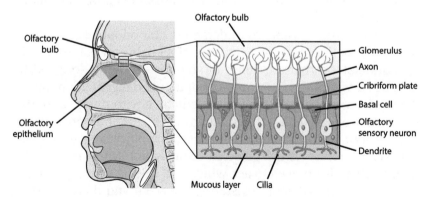

Figure 7.1 The process of olfaction (© Shutterstock).

7.2 ODOUR RECOGNITION (BOX 7.1)

> **BOX 7.1   DID YOU KNOW?**
>
> It is thought that we can identify up to 10 000 different smells, yet there are only a few hundred different types of olfactory sensory neurons that help to detect these different smells. How is this possible?

7.2.1   Specialized Olfactory Sensory Neurons

Axel and Buck won the Nobel Prize in Physiology or Medicine in 2004 for their discovery of odorant receptors and the organization of the olfactory system. They found 1000 different genes that code for olfactory receptors in the mammalian genome, giving rise to 1000 olfactory receptor types. They showed that olfactory sensory neurons possess only one type of odorant receptor, which can detect a limited number of odour molecules. As a result, olfactory sensory neurons are highly specialized for a few odorant molecules.

7.2.2   Infinite Combinations

Most odours are composed of a unique combination of odorant molecules, so any given odour activates a combination of several odorant receptors. This leads to a combinatorial code forming an 'odorant pattern'. We can recognize many different odours owing to the near-infinite number of possible combinations of odorant molecules.

It is therefore possible to understand that odours exist only in the brain, as the result of a number of signals and patterns. It is also understood that our sense of smell is primarily designed to help us survive and thus warn us of potential dangers. An odour is detected as a change in the immediate environment and the emergence of a potential threat. At any time, our brain processes a huge amount of information with the same objective – to help us survive. Therefore, if the detected scent is not synonymous with danger, there is no need to act, which will free up capacity. After a while, the brain will stop registering the scent and therefore stop noticing its presence, commonly known as 'odour fatigue'.

7.3 SMELL AND EMOTIONS

So, what is our sense of smell useful for and why is it so important?

The olfactory and limbic areas in the temporal cortex depicted in Figure 7.2 are among the oldest parts of the mammalian brain. In a split second, odours can help us decide whether it is safe or not to approach or avoid certain situations as a matter of survival. Therefore, the close link between odours and emotion is evolutionarily adaptive. Pleasant or unpleasant odours result in approach or avoidance behaviours, which is supported by substantial research. Some studies have even shown that pleasant and unpleasant odours modulate the perceived pleasantness of male and female faces.

Research has also shown a strong link between our sense of smell and our emotions. Most odours are inherently affective,

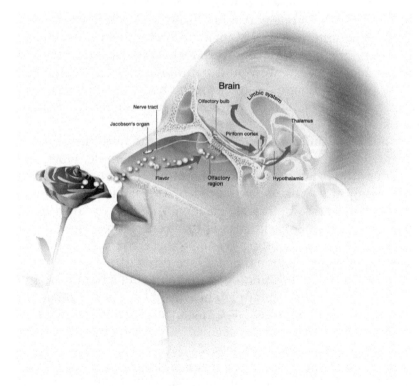

Figure 7.2 The olfactory bulb, limbic system (© Shutterstock).

which means that as soon as we smell a scent, we know if it is pleasant or not. Scents tend to provoke emotional reactions and trigger memories. This is due to the direct connections between the olfactory bulb in the brain and the limbic system. The limbic system contains the amygdala and the hippocampus, which are regions of the brain heavily involved in emotional processes.

7.4 THE MESSAGE CARRIED BY A FRAGRANCE

We now understand that fragrances are strongly linked to emotions. This is one of the reasons why cosmetic and personal care brands use fragrances in their products. Scent allows us to make a connection between the product we have elected to buy and its technical attributes, in addition to any claims made on the packaging. Fragrances are used to reinforce the marketing proposition in a non-visual way, helping the consumer to appreciate fully the product as a multisensorial experience.

It is important to note that many factors influence the choice of odours attributed to these products. First, let us consider the product category; we would naturally expect a shower gel, whose primary function is to clean the body, to smell different from a body moisturizer, whose key function is to nourish the skin.

In addition, different countries around the world have different expectations with regard to scents. For example, fabric detergent powders are fragranced differently in different countries, even though they may be produced by the same manufacturer and be from the same brand category. The primary function is the same everywhere and that is to clean clothes; however, a diverse history surrounding scent in these different countries strongly influences regional tastes (see Box 7.2).

Furthermore, the positioning in the market of a brand and the identity of the brand itself weigh heavily in the discussion about the type of scent required. For instance, fragrances suited to brands in the upper market segment are totally different from those of everyday basic products.

Professional and experienced teams in fragrance houses know exactly how to convey the right impression of a scent, in order to fulfil all of these criteria.

Follow the Scent – The Science Behind the Fragrance in Products 197

BOX 7.2 TASK

When you are in a different country, take the opportunity to visit a supermarket. Experience the types of scents in different products and question how they compare with the same products you use at home.

7.5 CREATING AND MASKING ODOURS

Scents punctuate our everyday lives: we wake up to the smell of our favourite shower gel or shampoo. We walk the streets to work on elegant clouds of fine fragrances and ride the day on clean whiffs of air fresheners, dishwashing liquids and washing powders. At the end of the day, we go to sleep with the reassuring aroma of applied face cream and body lotion.

Fragrances are incorporated into products to reinforce their technicality and enhance the experience. Large teams of trained professionals work together in 'fragrance houses' in order to create the best suited fragrances for our products. How are these stunning fragrances created?

7.5.1 The Creative Process and the Teams Involved

7.5.1.1 What is a Fragrance House?. The development of a fragrance is a complex creative process that involves teams of people. Many fragrances are born from a brief that is issued by a brand. The brief specifies the type of scent required for a new product or to extend the range of an existing products line. The brief is issued to fragrance houses, which create fragrances on behalf of a brand as an external supplier.

Let us examine how fragrance houses are structured (Figure 7.3).

7.5.1.2 Sales Teams. In a fragrance house, the account manager acts as the key contact for a brand. Their role is to gather the information required that helps to translate a written document. This brief conveys a particular message that also takes into account the colour of packaging and other visuals that may be used. The aim is to produce the best and most appropriate scent for the new product. It is crucial that everything is discussed with the client in detail, as 'quality in' translates to

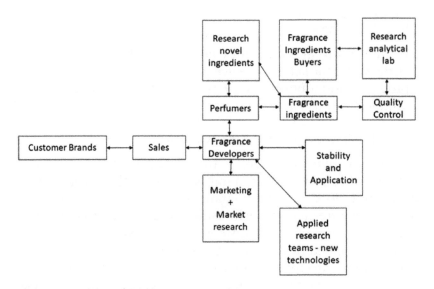

Figure 7.3 The fragrance house structure.

'quality out'. All relevant information is summarized in this internal document.

7.5.1.3 Marketing Teams. The marketing teams gather information on fragrance trends and follow the latest developments in specific areas of the market. Fragrance preferences are specific to geographical locations, brand values and product types. For example, a fragrance used in a shampoo is different from one used in a washing powder. Over time, fragrances also evolve according to the social environment and are influenced by many economic factors. Marketers needs to be able to analyse market research data; in that way they can best understand consumers' motivations and behaviours.

Marketing teams provide background information at the beginning of the brief, in order to set the creative teams on the right track. They also have a pivotal role once the fragrance has been created in order to validate it. Creations can be tested using market research, often with a group of carefully selected consumers being the target audience for the product.

7.5.1.4 Creative Teams. Creative teams in a fragrance house tend to be organized by category based on the end product

Follow the Scent – The Science Behind the Fragrance in Products 199

type: fine fragrances, personal care (shower gels, shampoos), fabric (detergents and conditioners), home care (dishwashing liquids, air fresheners, candles).

Perfumers and fragrance developers form the core of the creative teams in these houses. Their roles are complementary and together they develop the fragrance in answer to the brief from their client.

7.5.1.5 The Fragrance Developer's Expertise. Both the developer and perfumer will brainstorm their ideas together, imagining what the winning fragrance could be, possibly revisiting past creations as a good starting point.

To select fragrances with the best possible chance of being chosen by the client (as the overall winner), the developer will select against a number of rigorous criteria (Figure 7.4).

The performance of the product is based on immediate impact and 'substantivity' (the definition will be explained later in this section). Olfactive acceptance of the product is based on olfactive characteristics, product category, brand credentials, positioning in the market, target country, user and conformity to legislation/brand requirements. The developer evaluates a fragrance in its globality in order to meet all the requirements and not on the specific fragrance ingredients. The expertise of fragrance ingredient knowledge lies with the perfumer.

Figure 7.4 Summary of the fragrance developer's expertise; please note that neither personal preferences nor fragrance ingredients are part of this process.

For technical criteria such as stability and chemical interactions, the developer evaluates the fragranced samples after they have been exposed to temperature variations or light. They know how much variation is acceptable or not and decide if it is necessary for the perfumer to rework the fragrance.

Expert knowledge of the client's brand credentials is also a key factor. The developer can develop fragrances that are completely aligned with a brand's portfolio and will put aside their personal olfactive preferences in order to work objectively with the client.

After meeting with the developer, the perfumer works independently when creating the fragrances. These unique formulae are often created on computer, listing ingredients and precise dosages for each component while imagining the scent from memory. The perfumer only smells the creations once an assistant has weighted and mixed the fragrance ingredients together. Now the perfumer meets with the developer again to evaluate the creations together.

7.5.1.6 The Perfumer's Expertise. When creating a fragrance, the perfumer has many parameters to take in account (Figure 7.5):

- fragrance ingredients' stability and solubility in target base;
- costs constraints to meet the agreed price with the client;
- performance so that the fragrance can be perceived at the desired intensity in the final product;
- ingredient combinations to achieve the desired olfactive effect;
- ingredient combinations to achieve specific technical effects;
- regulation constraints that comply with the country of launch.

There are also a number of technical considerations that are specific to the product for which a fragrance is created. In the development of a shower gel, the perfumer will engineer the proportions of top, heart and base ingredients in order to cover the base odour of the product. It must provide 'bloom' in use and leave a long-lasting scent on the skin after rinsing. In some fragrance

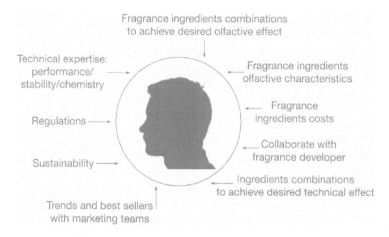

Figure 7.5 Summary of the perfumer's expertise.

houses, there is a wealth of data and software available for perfumers to support the optimization of these technical requirements.

It is therefore understandable that being a successful perfumer requires many years of training to master all these parameters, while at the same time creating beautiful fragrances (see Box 7.3).

We have established that the roles of both perfumers and developers are complementary. They work together in the development of the fragrance by an iterative process, where the developer will give feedback to the perfumer on the creations, until both parties are satisfied that the fragrance meets the required criteria. The fragrance will then be shared with the brand client, in answer to the brief.

7.5.2 The Construction of a Fragrance

The construction of a fragrance depends on the ingredients that enter its composition. A fragrance formula is very precise, yet it is possible to describe and classify a scent in different ways.

It is worth noting that no formal vocabulary has been established across the fragrance industry, therefore, terminologies can vary from person to person. There is, however, a general consensus towards a system that enables better communication about scents.

> **BOX 7.3 DID YOU KNOW?**
>
> On average it takes at least 10 years for a perfumer to master the art of creating fragrances.

7.5.2.1 Accords and Fragrances. What number of ingredients do you think constitute a fragrance? There is no strict rule to follow; however, the generally accepted norm is that if there are very few ingredients a composition is called an accord. Compositions that are more complex are considered fragrances.

An interesting question for discussion is 'What is the maximum number of ingredients a fragrance should have?' In theory, creations can have hundreds of ingredients. However, one should be careful with very complex formulas, for the following reasons:

- The more complex a formula, the more expensive it becomes.
- Even if your nose can distinguish a large number of materials individually, in a very complex formula some notes are lost and your nose cannot detect them. For this reason, some perfumers have taken to the habit of 'cleaning' their formulas once they feel that they have reached the desired olfactive character.

7.5.2.2 Pyramids and Volatility: Top, Heart and Base Notes. Aromatic ingredients in a composition have different specific physical and chemical properties. One of these parameters is volatility, which establishes the evaporation rate of substances or how much time they take to change from a liquid to a gas at room temperature and atmospheric pressure. Some aromatic materials will flash-off in minutes whereas others will linger for hours. This behaviour is linked to the molecular weight of the molecules of the ingredients: the higher the molecular weight, the less volatile they are, and *vice versa*.

To explain these differences in visual form, we can represent a fragrance as a pyramid. The most volatile ingredients are positioned at the very top, progressing towards less volatile components at the bottom. The different stages of evaporation (Figure 7.6) are as follows:

- Top notes take approximately 15–30 min to evaporate from the skin. Examples are citrus, aldehydic, green and marine notes.
- Heart notes take approximately 30 min to 4 h to evaporate from the skin. Examples are herbal, fruits, floral and spice notes.
- Base notes (or dry-down) take over 4 h to evaporate from the skin. Examples are woods, amber, leather, musk, resins, animalic and gourmand notes.

7.5.2.3 Fragrance Substantivity. From what we have learned so far, it is easier to understand how long a fragrance may be perceived and how long it might last on the skin. This is what we call the 'substantivity of a fragrance', which is determined by the volatility of its components (see Box 7.4).

If we take the example of a scent composed mainly of citrus notes, the scent will be very strong in the opening, but will last for only a very short time. This kind of fragrance is deliberately created for products such as dishwashing liquids, so that after cleaning no perceived odours are left on the dishes that may cause interaction with food.

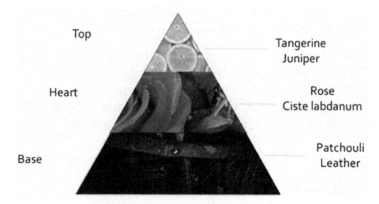

Figure 7.6 Fragrance pyramid.

> **BOX 7.4 TASK**
>
> To understand the concept of volatility better, you can follow the evolution of your favourite fine fragrance on your skin by recording your own description of the scent against time during the period of a day.
>
> Spray the fragrance on the inside area of your wrist. Do not spray the fragrance in your neck area as this will saturate your nose with the fragrance and will impair your ability to smell as part of a fair test.
>
> The most effective way to describe the perfume is to use your very own words, perhaps based on what the scent reminds you of and without being influenced by the marketing description or advertising.
>
> Make a note of both the time and the description of the scent at the start of the task. Keep a record of the scented notes that you detect on your skin at different time intervals during the day; first after 30 min, then at 1 h, 2 h, 4 h, 6 h and finally after 8 h. Try to approach each smelling session with an open mind and a fresh nose every time! Try to be as objective as possible without revisiting your earlier notes.
>
> At the end of the day, read through all of your comments and see how your descriptions changed over the time period. What did you find to be the most important characteristics for you personally after 2 h in comparison with those after 8 h?

In the case of a scent centred around base notes, such as woods and musk, for example, it will be difficult to pick up the odour when first applied. These notes become stronger over time and are very substantive. This is why these kinds of notes can be used in fabric detergents and softeners to achieve long-term substantivity.

7.5.3 Fragrance Families

We have just seen that fragrances can be described by listing the ingredients in order of volatility in a pyramid. Another way to describe a scent is to list the key accords, the predominant elements in the scent being listed first. In this case, we use the 'families' to classify the fragrances (see Box 7.5). Here are the key main olfactive families:

- citrus
- aldehydes
- green
- marine, aqueous
- fruits
- herbal, aromates
- floral
- spices
- woods
- leather
- musks
- powdery
- gourmand notes
- ambers
- animalic notes.

Three of the most common fragrance families have been formalized in perfumery – these are called fougères, chypre and oriental:

- *Fougères* – These are based on bergamot, lavender, coumarin and oakmoss.
- *Chypre* – These fragrances are based on accords comprising bergamot, geranium, ciste labdanum, oakmoss and patchouli.
- *Oriental* – These compositions include warm notes of vanilla spices, coumarin and soft woods and musks.

The main difference from the pyramid format of description is that the top notes of a fragrance might not necessarily be listed first if they are not the predominant notes in a composition. In the previous example, if the composition had more of a woody element, it would be described as a 'woods, citrus, aromatic' fragrance. If, however, the aromatic elements were predominant, it would be described as an 'aromatic, citrus woods' fragrance.

> **BOX 7.5 TASK**
>
> In order to understand better the concept of odour families, try to describe the different kinds of scented products you use. Some might be predominantly fruity (if so, try and challenge yourself to name the kind of fruit: is it a green fruit, such as an apple, or a red fruit, such as a raspberry or a blackcurrant? Others might be floral or woody. You will soon discover that certain types of products tend to be based around one specific family – see if you can work out which!

7.5.4 Fragrances in Different Bases and Products

Most of the products that we use, not just alcohol-based fine fragrances, are fragranced. So, how do the differences in the composition of various bases affect the perception of the fragrance in the end product?

Brands sometimes create an extended line of different products in keeping with the same olfactive theme. A good example is where a successful fine fragrance has been extended into a line of toiletries for gift sets containing body lotion, shower gel, deodorant and soap. This is to fulfil the consumers' wish to have a set of products where the scents are aligned and smell the same. For example, if the body lotion from the range is to retain the fine fragrance for longer on the skin, it is crucial that the fragrance impression in the various bases must also remain the same.

Any attempt to use the eau de toilette (EdT) fine fragrance oil in the body lotion base will be unsuccessful. The EdT fragrance will be distorted and hardly perceivable, because the alcohols used to carry fragrance and body lotion bases have very different physical and chemical properties.

Soaps, shower gels, creams and deodorant bases also all have very different properties. They require a differently engineered formula of the same olfactive theme in order to achieve a similar or comparable scent impression. The perfumer must therefore create different fragrances of a similar olfactive theme in order to give the same impression in the different bases.

Let's take a look at some of the chemistry to help us detail their differences further. In an EdT, ethanol forms the base of

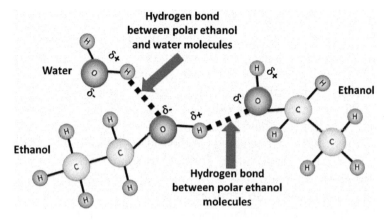

Figure 7.7 Ethanol molecules act as small magnets.

the product (Figure 7.7). It is described as polar because its formula contains a hydroxyl (OH) group. The hydroxyl group attracts and creates bonds with other polar molecules by acting like a small magnet. Ethanol will therefore 'like' other polar molecules, which explains why polar ingredients will dissolve in ethanol. Non-polar ingredients will not dissolve, which explains why sometimes small drops of oil can form in the mixture.

The volatility of alcohols is high, which encourages evaporation of the fragrance ingredients. This explains why ethanol-based products tend to magnify fragrances very effectively.

In a body lotion, the base is formed of a mixture of water with oil and surfactants to create an emulsion. Water is a very polar substance, whereas oils are non-polar. Hence the body lotion will 'like' both polar aromatic fragrance molecules, thanks to water content, and non-polar aromatic molecules, thanks to the oil.

We should also remember that water and oil are less volatile than ethanol. The combination of both the polarity and volatility differences with ethanol explains why in general a fragrance appears to have much less impact in a body lotion than in ethanol.

In order to obtain the same olfactive impression in an EdT and a body lotion, the perfumer needs to create different fragrance versions of the same olfactive theme.

7.5.5 Covering Malodours

Historically, fragrances have been used to cover bad smells. One of the earliest examples was in the making of basic soaps using wood ash lye (wood ashes and water) mixed with fat. Depending on the geographical location, the fat could have been of vegetable origin (such as olive oil used in Marseilles soap) or animal origin, which might have been more readily available. Animal fat in soap would have caused undesirable malodours that fragrances would need to mask. By trial, error and experimentation, perfumers soon realized that some fragrance ingredients were better at covering these 'animalic' notes.

Over the years, extensive research and development have enabled perfumers to engineer their fragrances very precisely to optimize the desired effect. This is to cover not only undesirable product bases as described previously, but also everyday undesirable smells, from toilets and kitchens and even unpleasant body odours. Today, fragrance houses use technology and hold extensive data. They have a wealth of knowledge about how to design a fragrance to disguise a range of malodours. This encompasses not only ingredients that are the most effective at counteracting a specific off-note, but also the precise combination of ingredients to use.

7.6 STABILITY: WHY DO FRAGRANCES CHANGE OVER TIME?

7.6.1 Base Interaction

In addition to widening a perfumer's palette and allowing them to save on natural resources, aroma chemicals have also enabled perfumers to create scents for difficult bases. Challenging bases such as acidic bases for fabric conditioners or alkaline bases in bleach will degrade natural extracts, leading to discoloration and unwanted harsh odours.

A natural extract is composed of a wide range of naturally occurring chemicals, with very different properties. Lavender oil can contain up to 100 ingredients, such as alcohols, esters, ketones and terpenes. Each of these ingredients contributes intricately to the scent of lavender; they are stable in some bases but degrade in others. It is not possible to separate the components of a natural extract, otherwise the scent would no longer be that of lavender and would have a different chemical profile. Thus, using natural extracts in certain bases can lead to unstable fragrances. In these cases, the use of aroma chemicals can often provide the solution. These are substances with minimal impurities and they have only one chemical functional group with a very specific reactivity. The perfumer can select the precise aroma chemicals to prevent reaction with the rest of the product components such as bases or packaging in a negative way. In other words, the perfumer selects the aroma chemicals that are compatible with the rest of the product formula.

7.6.2 Stability

We have seen that perfumers carefully choose the ingredients to compose a fragrance so that the scent is stable in the base for which it is designed. However, the truth is that over long periods of time, a degree of change is unavoidable. Understanding of the causes of these changes is attributed to a range of factors:

- Oxidation of certain ingredients by temperature or light and sometimes both (*e.g.* citrus ingredients).
- Ingredients reacting with each other induced by temperature, humidity or light (*e.g.* Schiff's bases in between anthranilates and aldehydes).
- Ingredients reacting with bases induced by temperature, humidity or light (*e.g.* ketones react with alcohol in EdTs to form ethanal).
- Discolouration caused by temperature, humidity or light of certain ingredients (*e.g.* indole, vanilla or vanillin).

Fragrance stability over time can be influenced by the base, but in addition the effect of packaging should not be underestimated. Glass is generally inert, but plastic containers can be permeable to certain fragrance ingredients, such as limonene. The reason is that the limonene molecule is quite small and has an affinity for certain types of plastic because of its non-polar-like polymer chains.

It appears that all elements of the product can influence the stability of the fragrance, as it is a highly complex system. Understanding the details of the chemistry involved in these systems can be long and difficult, hence protocols have been designed to predict at best the stability of a fragrance in a given base and its packaging. In these tests, great care has to be taken to separate the influence of each factor, in order to understand which remedy to apply to a specific issue. Temperature is tested without the influence of light or humidity, therefore in the dark and at constant humidity. Light is tested without the influence of temperature or humidity, therefore at constant temperature and constant humidity. Humidity is tested without the influence of temperature or light, therefore at constant temperature and in the dark.

7.6.2.1 Assessing Stability – Temperature. The test protocols for temperature rely on the Arrhenius equation, which states that for every 10 °C increase in temperature a reaction rate doubles, which also means that the time to complete degradation is reduced by half.

Fragrances are designed to be stable for 12 months at 20 °C. It should therefore be possible to run accelerated tests at 40 °C and predict the stability over the course of 3 months. At high temperatures, however, some reactions might be triggered that would not happen otherwise, so products are also tested over a range of lower temperatures – this has led to the widely accepted fragrance industry tests of 12 weeks in the dark at 3 different temperatures which are 40°C, 20°C and 4°C.

7.6.2.2 Assessing Stability – Light. The sensitivity of a product to light is generally tested in a temperature-controlled light cabinet. This is one way to avoid the influence of temperature. A windowsill will be subject to temperature variations that will invalidate the results because they will not be reproducible.

A few hours of testing in a UV cabinet results in the same amount of UV light to which a product would be exposed in daylight over a period of 3 months.

7.6.2.3 Assessing Stability – Humidity. For certain types of product, it is essential to test the influence of humidity. This is the case for powdered products in cardboard boxes, such as laundry powder or even eyeshadow. These will be tested under humidity stress, in order to reproduce the extreme conditions

Follow the Scent – The Science Behind the Fragrance in Products

that can arise when being transported around the world. The tests are generally carried out at 40 °C and 80% humidity.

7.6.2.4 Assessing Stability (see Box 7.6). The experts in fragrance houses assess the products over time and while they are undergoing the stability testing. If a product develops signs of discolouration or malodour, then the perfumer will be notified.

Upon receiving the stability test results, the perfumer will have to modify the fragrance in order to avoid any negative impact at a later stage. New fragrances will be created and tested in exactly the same way until the fragrance in question behaves in an acceptable way.

BOX 7.6   TASK

Why not conduct a simple stability test of your own at home? You can use bergamot essential oil as an example. Divide equal amounts of the oil in small glass containers or tubes. Keep all parameters the same *i.e.* the same amount of product and type of container. Using opaque tape, cover the outside of container one so that UV radiation from daylight cannot affect the product. Place an uncovered container in the refrigerator in the dark and a covered container on a north-facing windowsill (avoiding direct sunlight). Make a note of the day and record an original description of the scent. Keep recording the scented notes of the windowsill sample in comparison with the sample in the refrigerator and continue over the course of one month, making weekly notes. Once again, try to evaluate the scents with a fresh nose and an open mind every time (without revisiting your earlier notes), so that you are as objective as possible.

At the end of the experiment, you can read all of the notes you have made and see how your descriptions changed over time. What did you find the key differences were after 4 weeks? This experiment highlights the influence of temperature (around 20 °C) *versus* a reference sample in the dark at 4 °C. In industry conditions, temperature would be tested in an oven at constant temperature and in the dark to avoid the influence of UV radiation.

Should the fragrance pass these tests, only then is it considered to be ready for production before being launched safely.

7.6.2.5 What Is the Best Way to Preserve a Fragrance?. Fragrances created by expert perfumers are designed to keep at their best for at least 1 year; however, it is advisable to store them in the best possible conditions to aid the preservation period. This involves avoiding variations of temperature and exposure to light in the case of products in transparent bottles, and for powdered products it is best to avoid high-humidity environments where possible.

7.7 ESSENTIAL CHEMISTRY

Until the beginning of the nineteenth century, perfumery was very much considered to be a craft. Perfumers were trained chemists and pharmacists who were able to extract aromatic substances from plants and incorporate them into special pomades that were popular at the time.

After the French Revolution and towards the end of the eighteenth century, a new class of consumer emerged in this newly stabilized French society: the affluent and fashionable middle classes. In England, under the reign of Queen Victoria, the wearing of perfume remained strictly controlled, with only a discrete dab being the acceptable norm. Wearing too much perfume was like wearing too much rouge, which was considered unsuitable for a lady.

After the death of Queen Victoria, there was much change in society, which included a demand for and consumption of fragrances. A new opportunity emerged that prompted perfumers to develop new scents to suit the tastes of the day. This surge in demand spawned the first major perfume houses and in the course of the second half of the nineteenth century perfumery took a new path in the development of new aroma chemicals.

7.7.1 How Were Aroma Chemicals Discovered?

Aroma chemicals were originally discovered by chance while developing new processes. This was true in the case of Baur in 1888, who discovered nitro musk while researching explosives.

Over time, research for scented molecules became more structured and aroma chemicals were discovered by analysing natural plant extracts, determining the structure of these naturally occurring chemicals and then finding ways to synthesize them. For example, coumarin was originally isolated from tonka beans in 1820. A few of the early discoveries are listed in Table 7.1.

These developments meant that perfumers would not have to rely solely on the plant as a source to obtain the ingredient,

Table 7.1 List of early discoveries of aromatic chemicals.

Year	Aroma chemical	Chemical formula	Description
1868	Coumarin		Coumarin smells sweet, vanilla like, a little reminiscent of hay with a slight powdery almond note. It occurs naturally in tonka beans, which are the pip or stone of the fruits from the tonka tree. Coumarin can also be found in lavender and dried hay. The chemical process for coumarin from synthetic origins was discovered by William Henry Perkin, an English chemist, in 1868. White coumarin crystals were first used in a perfume in 1882, by perfumer Paul Parquet, in Fougère Royale by Houbigant. This was the first time a synthetic ingredient was used in a fine fragrance. Its success was such that it spawned the fougères fragrance trend and family
1869	Heliotropine		Heliotropine is reminiscent of vanilla or almond and brings a balsamic character with powdery, floral aspects. Despite being similar to the scent of the heliotrope flower, heliotropine cannot be extracted from the flower because it does not occur naturally in the flower. Heliotropine was discovered in 1869 by Fittig and Mielk

Table 7.1 (*Continued*)

Year	Aroma chemical	Chemical formula	Description
1872	Citronellal		Citronellal is the main component in citronella oil and gives the oil its distinctive lemon scent. It was discovered by Dodge in 1872
1874	Vanillin		In 1858, Nicolas-Théodore Gobley first isolated vanillin as a pure substance from a natural vanilla extract. In 1874, Ferdinand Tiemann and Wilhelm Haarmann deduced the chemical structure of vanillin and developed a synthesis route from pine bark
1888	Musk xylene		Albert Baur accidentally discovered nitro musk while trying to produce trinitrotoluene (TNT) for explosives in 1888
1898	Ionones		Ionones are found in a number of essential oils, such as rose oil. They are part of a group of compounds known as rose ketones. β-Ionone is a significant contributor to the aroma of roses, while a combination of α- and β-ionone is characteristic of the scent of violets. They are derived from the degradation of carotenoids

which led to more preservation of a number of plant and animal species. In addition, aroma chemicals were much cheaper to produce than natural extracts.

These synthetic counterparts have enabled perfumers to broaden their olfactory palette and feed their creativity. The re-creation of the scent of violets and most fruits was mainly possible because of the use of aroma chemicals.

> **BOX 7.7 DID YOU KNOW?**
>
> The International Nomenclature of Cosmetic Ingredients (INCI) is a systematic naming system for all ingredients found listed on the packaging of the products we buy.

At the end of the nineteenth century, a growing demand for fragrances became more apparent, and the fragrance houses of Yardley and Crown both exported scented products to the rest of the world. This demand encouraged perfumers to seek higher volume production techniques, to which aroma chemicals were better suited to than natural extracts. This in turn allowed perfumery to reach an even wider audience and set the scene for the explosive growth of perfumery in the twentieth century.

To discover more about how the fragrances are named on the labels of your products, see Boxes 7.7 and 7.8. For an excellent overview of the important dates of discoveries in aroma chemicals, see *The Chemistry of Fragrances* by Charles Sell in the Further Reading list.

7.7.2 Categories of Aroma Molecules

Analysis of the structure of existing aroma chemicals has helped chemists to devise ways of creating new olfactive molecules. They have studied the relationship between the functional groups and the odours.

Some of the most common perfume ingredients are included within the oxygenated monoterpenoids, containing one or more oxygen atoms. They are popular to use because they have the required optimum volatility that allows us to perceive them and smell them. Conversely, monoterpenes are hydrocarbons that do not contain oxygen, they tend to have weaker odours and so are of less interest. Similarly, larger terpenes containing more than

> **BOX 7.8 TASK**
>
> You might be interested in spotting some aroma chemicals in the INCI lists of the products you use in your home. See how many you can find.

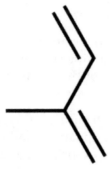

Figure 7.8 Isoprene unit structure.

two isoprene units are less volatile, meaning that there are fewer present in the air and so they are more difficult to detect.

Terpenoids can be visualized by linking isoprene units 'head-to-tail' next to each other (Figure 7.8). The arrangements can be linear or cyclic. Examples of terpenes and terpenoids are given in Table 7.2.

7.8 EXTRACTION METHODS

Chemists originally took inspiration from the natural world to create new aromatic molecules. In order to do so, they devised clever extractions methods that have since evolved over the years. As research advanced and new technologies developed, three main extractions methods have perdured:

- expression
- distillation
- solvent extraction.

7.8.1 Expression

Also called cold-pressed expression, this method is widely used for citrus fruits. Here the peel of the fruit is rolled over a trough. In this method (Figure 7.9), the tiny pouches in the skin (which contain the essential oil) are pierced. The process is carried out under running water, hence creating an emulsion with the less polar aromatic oils. The emulsion is then centrifuged to separate the two phases, leaving the aromatic oil floating on top. Extracts

Table 7.2 Examples of terpenes and terpenoids.

Category	Comments	Chemical formula	Plant source
Linear terpenes	These are composed of isoprene units arranged in a linear configuration. Myrcene is a linear monoterpene composed of two isoprene units and is the simplest terpene. It smells of woods and spices and can be naturally found in many plants, such as thyme	Myrcene	Thyme © Shutterstock
Cyclic terpenes	Cyclic terpenes, such as limonene, are composed of isoprene units arranged in a cyclic configuration. Limonene is a citrus ingredient reminiscent of oranges. It can be found naturally in orange and lemon skins	Limonene	Orange © Shutterstock
Oxygenated terpenoids: alcohols	These alcohols can be linear or cyclic, *e.g.* menthol, and contain one or more OH groups. Menthol smells aromatic minty. It can be found naturally in mint	Menthol	Mint © Shutterstock

Table 7.2 (*Continued*)

Category	Comments	Chemical formula	Plant source
Oxygenated terpenoids: aldehydes	These aldehydes can be linear, *e.g.* citronellal, or cyclic, and contain one or more C(O)H groups. Citronellal smells of citrus, reminiscent of citronella. It can be found naturally in lemongrass	Citronellal	Lemongrass © Shutterstock
Esters	Esters can be linear, such as geranyl acetate, or cyclic, and contain an $R^1C(O)OR^2$ group. Most esters exhibit fruit odours. Geranyl acetate smells of fruits and rose. It can be found naturally in roses	Geranyl acetate	Rose © Shutterstock
Ketones	Ketones can be linear or cyclic, *e.g.* α-methylionone, which contains an $R^1C(O)R^2$ group. α-Methylionone smells floral, reminiscent of violet. It can be found naturally in violets but not extracted	α-Methylionone	Violet © Shutterstock

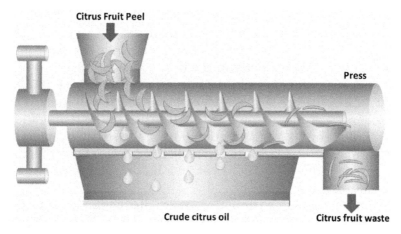

Figure 7.9 Cold-pressed expression.

BOX 7.9 TASK

In order to find out how much oil is contained in the skin of citrus fruits, squeeze the skin of an orange or lemon against a dry tissue paper. Hold the paper against the light to observe the areas that are marked. Experience the scent of the tissue and record how long it lasts during the day.

---

obtained by expression are called oils, *e.g.* lemon oil or bergamot oil (see Box 7.9).

7.8.2 Distillation

Distillation is primarily used to separate the components of a complex mixture into pure substances based on their different boiling points. The method consists in slowly heating a mixture and letting each substance evaporate individually. This happens because when the boiling point of a component is reached, only this substance will evaporate at that specific temperature, while the others will remain in the liquid state in the mixture until their boiling points are reached.

The distillation method in perfumery consists in heating plants to remove their aromatic components. This can be done by heating the plant directly, such as in the case of woods where the aromatic substances are less volatile and require higher temperatures. This process of distillation is called 'dry distillation'; however, high temperatures can lead to tar or burnt note elements. In the case of birch and cade oil, these notes are also part of the overall character of the oils.

In order to limit the temperature to a maximum of 100 °C, water or steam can be used to extract natural materials. The other benefit of this method is the decrease in the evaporation temperatures of the aromatic compounds, as they are swept away by the force of the steam and are co-distilled. This allows them to be extracted from the plant below their boiling points and limits degradation of the material.

Figure 7.10 shows a simplified steam distillation unit. In the receiving flask, the aromatic substances are separated from condensed water:

- Aromatic components are hydrophobic oils that separate from water at room temperature.
- Aromatic components have a lower density, which means that the oil phase floats on top of the aqueous phase.

The water can be recycled into the distillation process rather than being discarded as an effluent.

This method is well adapted to aromatic leaves, herbs, spices and certain flowers, such as rose, neroli and ylang ylang, that can withstand a temperature of 100 °C. Aromatic oils obtained by steam distillation are called essential oils, therefore we refer to rosemary, rose, neroli and ylang ylang oils as essential oils.

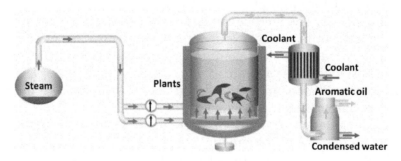

Figure 7.10 Simple steam distillation unit.

For plants that are heat sensitive, another method of extraction, solvent extraction, was developed.

7.8.3 Solvent Extraction

Originally this method was derived from enfleurage, a process in which flowers were left to soak in a bed of purified fat and then extracted with ethanol. Nowadays, enfleurage is considered too time consuming to be economically viable and solvent extraction has taken the lead.

Solvent extraction is similar to steam distillation, with two main differences: water is replaced with a different solvent and the process involves a second step.

First, let us examine different suitable solvents; examples are acetone with a boiling point of 56 °C and ethyl acetate with a boiling point of 77 °C, and they can be used as pure substances or in combinations to optimize the process. The common characteristic that they share is that their boiling point is lower than 100 °C, which allows the distillation of plants at lower temperatures, therefore avoiding damaging them.

As mentioned, there is a second step in the solvent distillation process, which is necessary because the solvent extraction of a plant produces a 'concrete' that is a solid that contains residues of the plant. This 'concrete' needs to be further refined by ethanol extraction and the ethanol is then distilled off to produce an 'absolute'.

This process is best suited to natural materials that are more sensitive to heat, such as many flowers. The resulting extracts are called absolutes, examples being jasmine, tuberose, orange flower and rose absolute.

Some plants can be extracted by both steam distillation and solvent extraction, which explains why you can find them listed in both categories; essential oil and absolute, as is the case with rose. Rose absolutes and essential oils will, however, display different olfactive characteristics and the perfumer will select one or another depending on the desired effect in the fragrance.

More recently, a solvent extraction method using supercritical carbon dioxide has been developed. In this case, the solvent is carbon dioxide (CO_2) in the liquid state or in a state between

> **BOX 7.10 DID YOU KNOW?**
>
> The most expensive extraction process is based on a gas that you can find in air – carbon dioxide, CO_2!

liquid and gas, thanks to specific temperature and high-pressure conditions. This method has several advantages:

- The extraction can be carried out at very low temperatures and so preserve the most volatile components of a natural material. This is particularly suited to ingredients that have very sensitive top notes, such as fresh ginger and pink peppercorn.
- There is no second-stage distillation to remove the solvent, as releasing the pressure will cause CO_2 to evaporate at 20 °C, resulting in no solvent residues.

The equipment required to sustain the extreme pressure conditions is expensive, which has limited the use of this method and its application in the field. It also explains why CO_2 extracts are expensive materials (see Box 7.10).

The most common perfumery extraction methods can be summarized as follows:

Natural material	→	Expression	→	Oil
	→	Steam distillation	→	Essential oil
	→	Solvent extraction	→	Absolute

7.8.4 Developments in Scientific Research Methods to Analyse Natural Scents

Analytical methods were originally based on extracting plants before analysing the extracts with the help of gas chromatography. This method was very efficient, but in the case of rare and endangered plants it was not suitable because the aromatic parts had to be cut away from the plant. For this reason, new methods were developed.

In the early 1970s, a new analytical technique was developed called 'headspace analysis'. This technology was exciting because

Figure 7.11 Equipment designed for non-invasive headspace analysis. (Image used with permission © Givaudan.)

it was non-invasive and focused on analysing the air surrounding the plant. A globe-like apparatus is placed above and around the plant, trapping the air that surrounds it (see Figure 7.11). The trap containing the air is then frozen and transported back to the laboratory. It is subsequently heated or rinsed with alcohol in order to retrieve the extract that is to be analysed with the help of gas chromatography. Today, headspace analysis is widely used to study endangered plant species around the world, leaving them intact and preserved.

7.9 CONCLUSION

We have only just started to unveil the richness and diversity of the world behind the creation of fragrances. Research is one of the key players in this discipline, from starting with understanding the mechanisms of our sense of smell to the discovery of aroma chemicals and new scented molecules. Essentially, creation is at the core of fragrance development, where ingenious perfumers skilfully translate complex and numerous data received from the developers, research, marketing and sales teams and turn this information into harmonious scents that it is hoped will appeal to the consumer.

Many people join the fragrance industry from different horizons, bringing with them diverse insights from other areas of expertise. Yet very few ever leave this fascinating world, as they soon become passionate about their work – the fragrance industry generates a powerful magnetic field that is hard to resist.

Ultimately, everyone involved, from the scientists to the creative teams collaborating with sales and marketing professionals, dreams of creating a fragrance that will capture the imagination of millions of people and mark their memory by creating a unique and iconic fragrance. When they succeed, the delicate balance of skills and collaboration with a complex network of professionals feels almost magical. Yet the amount of work and data accumulated over the years should not be underestimated and success has nothing to do with luck – remember it is the result of great creativity, precision work, well-coordinated disciplines and cutting-edge science!

FURTHER READING

Linda Buck and Richard Axel won the Nobel Prize in Physiology or Medicine in 2004 for their work on how the brain recognizes, categorizes and memorizes smell, You can read more details on the odour mechanisms from the 2004 Nobel Lecture at https://pdfs.semanticscholar.org/a886/d52c88540786ae04f6ddf1000f24319d4194.pdf.

S. Cook, N. Fallon, H. Wright, A. Thomas, T. Giesbrecht, M. Field and A. Stancak, Pleasant and unpleasant odours influence hedonic evaluations of human faces: an event-related potential study, *Frontiers in Human Neuroscience*, 2015, **9**, 661.

S. Cook, K. Kokmotou, V. Soto, H. Wright, N. Fallon, A. Thomas, T. Giesbrecht, M. Field and A. Stancak, Simultaneous odour–face presentation strengthens hedonic evaluations and event-related potential responses influenced by unpleasant odour, *Neuroscience Letters*, 2018, **672**, 22–27.

C. S. Sell, *The Chemistry of Fragrances: From Perfumer to Consumer*, Royal Society of Chemistry, 2006.

CHAPTER 8

The Inside Story – The Science Behind Active Ingredients

C. METCALFE AND T. CAUSER*

Adina Cosmetic Ingredients Ltd, Unit 8 Decimus Park, Kingstanding Way, Tunbridge Wells TN2 3GP, UK
*Email: tony.causer@adina.co.uk

If you look at the back of any cosmetic product, you will see a long list of ingredients that it contains, all of which have a specific function to ensure that the product is safe and effective. Previous chapters have looked at some of the different types of ingredients that are used in cosmetic products, such as emollients, rheology modifiers and fragrances. This chapter covers a wide range of active ingredients and product types, demonstrating how specific ingredients work in your products and why they are used. The chapter takes a look both at ingredients that are widely used in products to deliver 'everyday' functions, such as antioxidants and preservatives, and at some of the industry's 'special' palette of active ingredients, such as peptides and vitamins – these are the ingredients that brands like to shout about on the packaging. Although this chapter looks at just a few of the active ingredients typically used (those more commonly found in skincare products),

> **BOX 8.1 WHY IS WATER USED IN SO MANY COSMETIC PRODUCTS?**
>
> Water is used in many types of cosmetic and personal care products, such as lotions, bath and shower products, cleansing products, makeup and oralcare products. It is commonly used as a solvent, which means that it dissolves many of the other ingredients that are included in the formulation to perform specific functions in the formulation. When mixed with oil (and an emulsifier) it forms an emulsion, a format common to products such as skin and face creams.

every ingredient used in a cosmetic formulation is added for a specific function. You might want to peruse the ingredient list on some of the products you have at home and compare which ingredients are common across a number of products. Even water is included for a reason! (See Box 8.1).

8.1 VITAMINS

We are acutely aware that vitamins are vital to our bodies when consumed as part of our diet, but they can also add great benefits when applied topically to the skin or even to protect the integrity of the product itself. The way in which vitamins interact with the skin is key to the protection and benefits that they provide. This chapter will discuss both those which are oil soluble, such as vitamins A and E, and those which are water soluble, such as vitamins C, B_3 (niacin) and B_5 (pantothenic acid). All of these are commonly used in cosmetic products. Have you ever wondered where vitamins come from or why they are included in your cosmetic products? Read on!

8.1.1 Where Do Vitamins Come From?

It is a common misconception, and it is widely assumed, that vitamins in cosmetics are of natural origin. Although this is true of vitamins found in fruits and vegetables consumed as part of a healthy and balanced diet, supplementation of vitamins for oral intake and those found in skincare products for application to the skin are usually synthetic. There are a number of reasons for this, but key factors are the stability of the vitamins

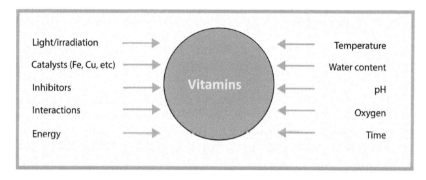

Figure 8.1 Factors that affect the stability of vitamins.

(see Figure 8.1) and the purity. Vitamins of natural origin, once exposed to air, temperature and light, degrade very quickly, leaving them less potent and effective.

8.1.2 Oil-soluble Vitamins

Oil-soluble vitamins, such as vitamins E and A, provide a wide range of anti-ageing benefits, much greater than their water-soluble counterparts because of their affinity with the lipids (fatty materials) found in the skin. As you will have read in Chapter 5, these lipids help to 'glue' the skin cells together.

8.1.2.1 Vitamin E: A Staple Ingredient. The term 'vitamin E' covers a group of eight fat-soluble compounds that are found in everyday oils such as olive and sunflower: four tocopherols and four tocotrienols (see Box 8.2). The structure of tocopherols has an impact on how the body and skin interact with them and how our bodies use them. One tocopherol, α-tocopherol, has the greatest 'bioavailability', meaning that it is very easily absorbed and used by the skin and has a greater presence than the other tocopherols. α-Tocopherol is also the

---

BOX 8.2   TOCOPHEROL AND TOCOTRIENOL

Tocopherol and tocotrienol are the names used to identify different structures of otherwise similar chemicals. The difference between these two may impact how well each is absorbed by the skin.

> **BOX 8.3 DID YOU KNOW?**
>
> Vitamin E is such a staple ingredient in personal care products that it is the world's most used cosmetic active ingredient!

main dietary source found in the European diet because of its presence in olive and sunflower oils.

Vitamin E protects cell membranes in our skin by stopping the production of free radicals or 'peroxides', which attack and break down our cell membranes, resulting in ageing of the skin and a compromise in the skin's natural barrier.

The form of vitamin E used in cosmetics, namely tocopheryl acetate, offers a high degree of protection for the product in addition to protecting the skin (see Box 8.3). Its ability to prevent the breakdown of oils (through oxidation) means that it helps preserve the life of oil-soluble ingredients, minimizing poor odour and maintaining the benefits that the oils have.

8.1.2.2 Vitamin A: Not Just Key to Human Sight. The term 'vitamin A' covers a group of molecules all derived from retinal, a chemical converted from β-carotene (the substance that makes carrots orange) (see Box 8.4).

The most common derivatives are retinol and retinyl palmitate, which are of most interest and use in the cosmetic industry. Retinyl palmitate is a reaction product (combining retinol and palmitic acid) and is much more stable and easier to formulate with.

Vitamin A can be obtained from animals or plants:

- Animal foods such as liver (see Box 8.5), fish liver oil, whole milk, cheese and whole eggs provide retinol or preformed vitamin A (active and readily available).
- Plant foods such as colourful fruits and vegetables provide the carotenoids that convert to vitamin A.

> **BOX 8.4 THE GROUP OF VITAMIN A MOLECULES**
>
> - Retinol
> - Retinyl esters (*e.g.* retinyl palmitate, retinyl stearate)
> - Retinoic acid

> **BOX 8.5 DID YOU KNOW?**
>
> Polar bear liver is so rich in vitamin A that eating it is fatal to humans. Arctic explorers found this out the hard way!

Retinol has long been viewed as a gold standard in anti-ageing. It is the conversion of retinol to retinoic acid in the skin with the help of enzymes that aids normal skin development. This is reported to help increase the collagen and elastin content of the skin, leading to a reduction in the appearance of fine lines and wrinkles and improving the firmness and elasticity of the skin.

Retinol is also known to reduce discolouration and improve skin tone. The melanin that provides (sometimes unwanted) colour to skin is decreased by minimizing its formation with retinol.

8.1.3 Water-soluble Vitamins

8.1.3.1 Vitamin C: A Natural Protector. Vitamin C is abundant in the skin as its own defence against oxidation (a form of damage) (see Box 8.6). You only have to see just how quickly an apple goes brown when cut and exposed to air to understand the impact of oxidation. You'll read more about oxidation later. As vitamin C is easily lost from the skin when exposed to stresses and sunlight, replacing the lost vitamin C can be hugely beneficial.

Sodium ascorbyl phosphate is a stable form of vitamin C (less affected by light and air) and on application to the skin is easily converted into L-ascorbic acid, the true form of vitamin C. It is this stabilized form that is widely used in cosmetics.

> **BOX 8.6 SUBSTANCES DERIVED FROM VITAMIN C**
>
> - Ascorbyl palmitate
> - Ascorbyl glucoside
> - Magnesium ascorbyl phosphate
> - Ascorbic acid

Vitamin C is well known for a variety of uses, such as skin lightening, anti-acne and enhancing skin firmness through stimulation of collagen production. Some forms of vitamin C have other specific uses: inactivating bacteria in the mouth when used in toothpaste, helping to protect the teeth and gums and acting as a deodorant by minimizing the growth of odour-causing bacteria under the arm.

8.1.3.2 Vitamin B: A Multi-functional Family. The family of B vitamins includes thiamin (B_1), riboflavin (B_2), niacin (B_3), panthothenic acid (B_5), pyridoxine (B_6) and biotin (B_7), with the most popular, in terms of application to the skin and hair, being niacin (in the form of niacinamide) and panthothenic acid (in the form of panthenol).

8.1.3.2.1 Niacinamide (Vitamin B_3). Niacinamide (vitamin B_3) is a very stable vitamin that offers a wide range of well-documented benefits when applied to the skin.

Vitamin B_3 protects UV-stressed skin by preventing pigments from surfacing on the skin. This lightens the skin and leads to a more even skin tone. It is also well known for reducing the formation of spots and pimples by targeting sebum production and inflammation. Recent studies have demonstrated that vitamin B_3 offers the skin a protective effect against urban particle pollution.

8.1.3.2.2 Panthenol (Vitamin B_5). Panthenol, a synthetic substance, is better described as a pro-vitamin. A pro-vitamin is converted into a vitamin through biological processes in the skin. Upon application of panthenol to the skin it bio-converts to pantothenic acid (vitamin B_5). Pro-vitamin B_5 (panthenol B_5) helps form part of the structure of coenzyme A, a vital substance in hundreds of body processes, some of which are key in the maintenance and repair of cells in the skin and hair.

Panthenol is valued for its moisturizing properties in both skincare and haircare applications, but it also helps to soothe irritated skin. It is hygroscopic (attracts moisture from the air) and adheres strongly to the hair, penetrating deep into the cortex of the hair shaft, helping to strengthen the hair and making it more resistant to breakage.

You may have seen products advertised as containing niacinamide or pro-vitamin B_5.

8.1.4 Minerals

Both vitamins and minerals are essential for the body. However, there are differences. Vitamins are organic (containing carbon and thus derived at some point from living organisms such as plants and animals) whereas minerals are inorganic (do not contain carbon). Minerals are not broken down by the body – in particular the skin – and are much less relevant to personal care products than vitamins.

8.1.5 Other Vitamins

This section has covered the main vitamins used in the cosmetic industry, but there are some missing. This is because some vitamins, such as vitamin K, are not allowed for use in cosmetics (at least in Europe) and others, such as folic acid, have limited recognized benefits if applied to the outside of the body.

8.2 PEPTIDES

A peptide is the middle child in a family of related molecules. It is smaller than a protein molecule but larger than an individual amino acid, but is made up of the same basic components: carbon, hydrogen, oxygen and nitrogen. Peptides contain between two and around 50 amino acids chemically joined together in chains, but most peptides found in the human body are typically at the shorter end of this range with chains of around 20 amino acids. They are naturally present in living organisms but can also be created synthetically either by building up chains of amino acids or by breaking down proteins.

There are about 500 naturally occurring amino acid building blocks and, with chain lengths of up to 50 amino acids, the number of variations possible is almost limitless. Each combination produces a specific effect. This is why the biological activity of peptides can be so varied.

Amino acid chains longer than 50 are known as proteins. Proteins are large, complex molecules with near-infinite

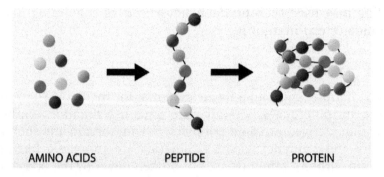

Figure 8.2 Proteins are molecules made up of many amino acids (© Nasky/Shutterstock).

combinations of amino acids that are integral to many of the structures in our bodies (see Figure 8.2). Their long chains fold into complex three-dimensional structures and they can be incorporated into cosmetics – typically for moisturizing or conditioning properties. They are too large to penetrate the skin and so are not very bioactive as they don't come into contact with the living parts of the skin beneath the surface.

The vast number of peptides in existence means that they are some of the most useful molecules in living organisms and we are only just beginning to understand their complex functions, interactions and involvement in communication inside and between the cells in the body. Naturally occurring peptides are signalling molecules – they tell cells when to build proteins, when to divide, when to relocate and when to produce substances such as proteins or pigments. This means that they are essential in many biological processes relating to the creation, maintenance, life and death of all the cells in the body.

8.2.1 Why Are Peptides Useful in Cosmetics?

Peptides are widely used in cosmetic products owing to their potential to modify how skin cells behave. They are often the silent workhorse behind effective skincare products. Targeted cosmetic peptides can be used to increase the production or reduce the breakdown of skin structural proteins (such as collagen or elastin), increase or reduce production of the skin pigment melanin (to brighten or darken the skin's complexion) or boost the skin's own antimicrobial defences. They can even be attached to other useful cosmetic components as a delivery system – to make sure that the active component reaches the correct target in the skin.

8.2.2 How Are Peptides Named?

You can generally spot a peptide on a cosmetic ingredients list because the name will contain the word 'peptide', often combined with the number of amino acids that make up the molecule. For example, a dipeptide (di = 2) will consist of two amino acids whereas a tripeptide (tri = 3) will consist of three (see Figure 8.3). Peptides are often then attached to a fatty acid group to make them more stable and help them to penetrate the skin. A common example you might see is palmitoyl tripeptide-1. In this molecule 'palmitoyl' is the fatty acid group and tripeptide lets you know that there are three amino acids in the chain.

8.2.3 Discovering New Peptide Ingredients

Peptides exist naturally in the skin as a result of either being naturally produced or the breakdown of larger protein molecules, but the peptides used in cosmetics are typically synthetic (manufactured). Natural peptides could be expensive and inefficient to produce and also are unlikely to penetrate the skin or may

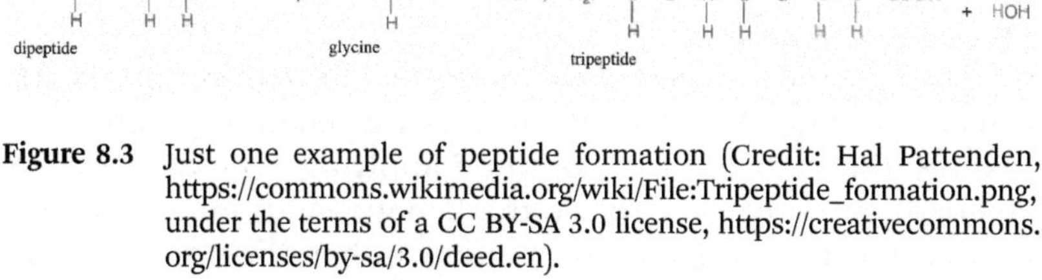

Figure 8.3 Just one example of peptide formation (Credit: Hal Pattenden, https://commons.wikimedia.org/wiki/File:Tripeptide_formation.png, under the terms of a CC BY-SA 3.0 license, https://creativecommons.org/licenses/by-sa/3.0/deed.en).

even be broken down by the biological processes of the skin before they reach their intended target. For this reason, peptides are typically synthetic although they are often designed to mimic the size, shape or amino acid content of the natural equivalent. Peptides that are believed to have biological activity (collagen boosting, for instance) can be quickly screened in a laboratory using cultured (manufactured) human skin cells to see which show the strongest effect. Once an active peptide has been identified, it will then have to go through further testing to ensure safety and stability and that the theoretical effect can be seen *in vivo* (on a human volunteer). They are scientifically advanced ingredients and the best ones will have undergone years of rigorous testing to confirm both their safety and their effectiveness.

8.2.4 Which Peptides Are Commonly Used?

A combination of the synthetic peptides palmitoyl tripeptide-1 and palmitoyl tetrapeptide-7 is one of the most common and best-known peptide blends. It is included in hundreds of cosmetic products owing to its reported ability to address the signs of ageing by boosting collagen and elastin production.

Acetyl hexapeptide-8 is commonly used in products targeting expression lines on the forehead and around the eyes. It has been designed to mimic a fragment of botulinum toxin (Botox®) and, if it can reach the relevant target in the skin, it might have a similar short-term effect.

Similarly, the dipeptide diaminobutyroyl benzylamide diacetate was designed to mimic waglerin-1, which is a peptide found in the venom of the temple viper (see Figure 8.4). Like acetyl hexapeptide-8, it is marketed for its ability to help relax expression wrinkles.

8.2.5 Is It Just Hype?

Peptides *can* be very effective, with many being shown in scientific studies and peer-reviewed literature to deliver significant cosmetic benefits. However, the effects are generally progressive, meaning that you need to continue applying a product for an extended period of time to see the benefit. Additionally, a peptide that has been broken down either in handling or transport

Figure 8.4 A temple viper (© fivespots/Shutterstock).

before it is applied to the skin, or has not been delivered effectively to the living target below the upper layers of the skin, will just be an expensive and elaborate moisturizer.

8.3 HYDROXY ACIDS

The α- and β-hydroxy acids are ingredients commonly used in skincare products to cleanse pores, smooth fine lines and improve skin condition through exfoliation (removal of dead skin cells). How much exfoliation or cleansing is achieved depends on the type and amount of the hydroxy acid and the pH of the product. Glycolic acid is the most commonly used α-hydroxy acid (AHA) (used for skin exfoliation) but, among others, lactic and citric acids are also commonly found in products. These common AHAs are derived from natural sources – sugar cane, milk and fruit. Glycolic acid can penetrate deep into the skin, making it a highly effective ingredient at tackling pigmentation caused by the Sun. The β-hydroxy acid (BHA) salicylic acid not only exfoliates the top layer of the skin but also helps to cleanse the skin by targeting deep inside the pores, helping to prevent the oily build-up that can lead to spots. Salicylic acid also has antibacterial properties. Selecting either an AHA or a BHA delivers similar yet different benefits in the product.

8.3.1 Why Are Hydroxy Acids Useful in Cosmetics?

You will have read in Chapter 5 that skin is constantly renewing itself and shedding old skin cells to replace them with new ones. Hydroxy acids are added to formulations to help speed up this epidermal renewal process. They act to exfoliate the skin by disrupting the 'glue' that holds the cells at the base of the stratum corneum together. This leads to old cells breaking off at a faster rate from the surface of the skin, producing a smoother looking skin surface and helping to reduce hyperpigmentation (excessive colouring). As cells are removed at a faster rate, new cells must be produced to replace those being lost. The new cells create the effect of 'younger looking skin' and a more even skin texture than you would see from old cells sitting on the surface waiting to be removed.

The level of AHAs included in the formulation will impact the effect seen on the top layer of the skin. As lower levels will produce a smaller exfoliating effect, younger skins that display fewer signs of ageing may only need to use products with lower levels of AHAs, *e.g.* <5%, whereas older skin may benefit from, and be able to tolerate, a higher level of AHAs, *e.g.* 10%.

As acids, hydroxy acids work best at a low pH. Products are formulated to contain hydroxy acids below a level that may cause irritation (usually <10% of the formulation for glycolic acid, 2% for salicylic acid) and with a low pH of 3–5 (acidic) in order to ensure that the product is gentle on the skin but still effective at exfoliating. Products that are left on the skin (*e.g.* moisturizing creams) will normally contain lower levels of AHAs than those which are rinsed or removed shortly after application. In the EU, AHAs and BHAs are tightly regulated by cosmetics legislation, particularly minimum pH requirements for AHAs and maximum use levels for BHAs.

8.3.2 Hydroxy Acids in Peel Products

Peel products will often contain high levels of AHAs (>20% of the formulation), commonly glycolic acid, and are designed to be used infrequently to remove the outer layer of the skin.

In some regions of the world, the stronger peel products that act at a deeper level of the skin are not considered to be cosmetic products and may be regulated as a different class of product.

Such products are often regulated as medicines or drugs because they target specific skin conditions such as acne or discolouration.

8.3.3 Using Products Containing AHAs

When using products that contain AHAs, it is generally recommended also to apply a sun protection product and avoid exposure to the sun for a week after use as AHAs increase the skin's sensitivity to UV light. The effect of the hydroxy acids on sun sensitivity is only short, however, and is fully reversed once the use of the AHA-containing product is discontinued.

8.4 UV FILTERS: PROTECTING PRODUCTS AND THE SKIN/HAIR

The Sun's ultraviolet (UV) and visible light rays can be damaging to the body and to cosmetic products.

8.4.1 The Electromagnetic Spectrum

Both UV and visible light are part of what is known as the electromagnetic spectrum (you read about this in Chapter 6). This also includes microwaves, radio waves, WiFi and X-rays. Within the spectrum, the ultraviolet rays sit close to the visible light rays and can be divided into three categories: UVC rays, which extend from 100 to 288 nm, UVB rays, extending from 288 to 320 nm, and UVA rays, extending from 320 to 400 nm.

The majority of UVC rays do not reach the Earth's surface from the sun, primarily due to filtration by the ozone layer, and therefore do not constitute a risk to the general population.

UVB rays penetrate the ozone layer and are responsible for most of the skin-related effects caused by exposure to the Sun. These rays can cause permanent changes to a person's genes, which is described by the term 'mutagenic'. The rays cause erythema (sunburn) – the reddening of the skin caused by the sun – and chronic UVB exposure is also the primary cause of skin cancers. However, UVB exposure can also provide beneficial effects such as pigment formation, which provides protection and production of vitamin D, important for processing calcium in the body and the production of serotonin, linked to happiness and serenity.

> **BOX 8.7 DID YOU KNOW?**
>
> UVB rays cannot penetrate glass, but UVA rays can. This means that whereas you cannot get sunburnt while travelling in a car, your skin will still suffer from the long-term effects of sun damage. This can be seen in people who spend a lot of time driving, when one side of their face can often look more 'weathered' than the other.

UVA rays cause an immediate reddening, skin darkening (known as persistent pigment darkening and melanin pigment formation) and colour changes to hair (see Box 8.7). These rays also damage skin collagen fibres and start an abnormal production of elastin. When this happens, we see poor skin repair and wrinkles. UVA causes damage to the DNA (genetic material) within cells deep within the skin. Repeated UVA exposure increases the risk of cancer and it can also impact the immune system.

8.4.2 What Is Sun Protection Factor (SPF)?

The sun protection factor (SPF) indicated on cosmetics was introduced in 1974 and is a measure of the fraction of sunburn-producing UVB rays that reach the skin. The SPF test consists of determining the amount of UV light that causes solar erythema on the area protected by a sunscreen compared with unprotected skin. The whole UV spectrum is used. Human volunteers are usually used for testing, but *in vitro* methods (tests that take place in test-tubes or other laboratory equipment) are being developed that will allow the SPF to be determined without exposing volunteers to UV light. The SPF test measures visible damage caused mostly by UVB rays, but invisible damage and skin ageing will also be caused by UVA rays during the test.

In the EU, sunscreen labels are only permitted up to SPF 50+. This is because 100% protection is not possible; a sun product with SPF 30 filters almost 97% of UVB rays and one with SPF 50 filters 98% of UVB rays. Health experts recommend a minimum of SPF 15, which filters 93% of UVB rays. The number indicates how many times longer it will take the skin to burn than when not wearing sunscreen. For example, if you typically burn after 10 min in the sun with no protection, applying an SPF 30 sunscreen means that you are protected for up to 30 times longer or

> **BOX 8.8 USING SUNSCREEN**
>
> Sunscreen should never be used so that you can stay in the sun longer. Always re-apply it regularly and especially after swimming.

> **BOX 8.9 UV PROTECTION**
>
> The most effective sunscreens protect against both UVB radiation, which can cause sunburn, and UVA radiation, which damages the skin with more long-term effects, such as premature skin ageing. Such suncare products are referred to as 'broad spectrum'.

300 min. Applying the sunscreen according to the instructions is integral to the protection it offers (see Box 8.8).

8.4.3 How Does UVA Protection Differ from SPF?

In addition to having SPF protection (protection against UVB rays), many products also offer protection against UVA rays (see Box 8.9). The UVA protection factor, UVA-PF, is established by determining the UVA dose that causes persistent pigment darkening, which results from the breakdown of melanin. In the EU, for a product to carry the UVA logo, the UVA-PF must be at least one-third of the SPF but a number is not indicated on the label. In the UK and Ireland, a star rating system is sometimes also used, with up to five stars on the label showing the level of UVA protection. The more stars, the higher is the protection. Five-star products offer the highest protection.

8.4.4 How Do Sunscreen Products Work?

Most sunscreen products work by containing either an organic compound (contains carbon) that *absorbs* ultraviolet light (such as phenylbenzimidazolesulfonic acid) or an inorganic mineral material (does not contain carbon) that primarily *reflects* light (such as titanium dioxide and zinc oxide), or a combination of both. In most countries, legislation controls which UV filters may be added to cosmetics in order to protect the skin. In some countries and regions, sunscreens are considered as medicines.

Historically, sunscreen products left a white film on the skin and had a sticky residue. Consumer demand for lightweight suncare products has meant that sunscreen formulations have become much more refined over the last decade. The whitening typical of inorganic filters in the last century has been improved by reducing the size of individual particles and coating the particles. Organic UV filters are only soluble in oil, hence the perceived greasiness of suncare products containing them. With the use of new ingredients that offer non-greasy and non-tacky feel, sunscreen products are becoming comparable to sophisticated skincare products but with the added benefits of a high SPF and UVA-PF.

8.4.5 Product Innovation

Consumers increasingly want products that are effective throughout the various situations of everyday life. These include swimming, contact with sand, perspiration, high heat and high humidity. Specific UV filters that form a film on the skin enable water resistance to be achieved.

8.4.6 Why Do Coloured Cosmetics Sometimes Contain UV Filters Even If They Don't Offer UV Protection?

Cosmetic products can also be affected by light, but UV filters can be used to protect them. Fragrances and dyes used in cosmetics are particularly vulnerable. As fragrances are typically packaged in transparent or translucent containers, UV filters either in the packaging or directly in the product offer protection. Indeed, various cosmetics require stabilization for both colour and fragrance to prevent instability caused by UV radiation and blends of UV filters such as octocrylene, homosalate and butylmethoxydibenzoylmethane are frequently used. You might also find these in your sunscreen products too!

8.4.7 The Hair Needs Protecting Too!

UV filters are used in some haircare products such as shampoos, conditioners and styling products to protect against the breakdown of hair proteins or colour loss. Currently, benzophenone and ethylhexyl methoxycinnamate (octinoxate) are the two UV filters most commonly used in hair products. The common UV filters used on skin are rarely used for hair products because they can leave the hair feeling heavy and unpleasant.

8.5 ANTIOXIDANTS

If there is such a thing as an 'antioxidant' then there must also be something called an 'oxidant'. To be able to understand what antioxidants are, we need to establish why we need them.

Terms such as 'free radical', 'antioxidant' and 'oxidative stress' have experienced almost explosive use by the media in recent years and raised enormous interest. This section is intended to provide an overview of oxidants and antioxidants in relation to the skin and their application in cosmetic products.

8.5.1 What Is Oxidation?

If you've ever had to deal with a rusty car or throw out browned fruit, then you have oxidation to blame. Oxidation can be a spontaneous process or it may be started artificially (see Box 8.10). Sometimes it is helpful and sometimes it is very destructive.

BOX 8.10 SOME EXAMPLES OF OXIDATION

Most metals react with oxygen to form compounds known as oxides. Rust is the name given to the oxide of iron and, sometimes, the oxides of other metals (see Figure 8.5). The process by which rusting occurs is also known as corrosion.

Figure 8.5 Rust (© Alberto Masnovo/Shutterstock).

It is the slow oxidation of oils and fats present in food material that results in some bad smells and taste (see Figure 8.6).

Browning of apples is a type of oxidation due to an enzyme present in the apples that reacts with oxygen when the apple is cut (see Figure 8.7).

Figure 8.6 Rancidity of food (© Nataly Studio/Shutterstock).

Figure 8.7 Fruit browning (© Darren Pullman/Shutterstock).

Fire is the rapid oxidation of a material and the process is called combustion, where the material reacts with oxygen releasing heat, light and various products (see Figure 8.8).

Figure 8.8 Fire (© Kompass/Shutterstock).

Oxidants are molecules that are missing one or more electrons. These oxidants are unstable and are constantly searching for their missing electrons. You may have heard of one example, 'free radicals', which can be damaging to the cells in our body. Antioxidants are 'hero' molecules that save the day by sacrificing themselves in order to protect our body's cells from attack (see Figure 8.9). Think of fish in the bottom of a fish tank, sucking up all the waste. Antioxidants are the molecular equivalent of free radical scavengers in the fish tank of your body.

8.5.1.1 What Are Free Radicals and Oxidative Stress?. Free radicals are molecules that can damage our body. For example, when our body uses oxygen as it does for breathing, it creates free radicals and the damage caused by those free radicals is called 'oxidative stress'.

The balance between oxidants and antioxidants is crucial for life. An imbalance will potentially lead to damage as oxidative stress. Oxidative stress can lead to diseases such as inflammatory

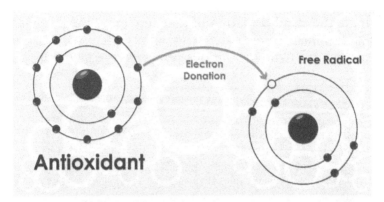

Figure 8.9 Antioxidants donate an electron to free radicals to stabilize them (© magic pictures/Shutterstock).

diseases, radiation damage, muscle and tissue degeneration, heart disease, diabetes, cancer and ageing of the skin.

8.5.2 How Do Antioxidants Work?

Antioxidants stabilize free radicals to minimize their harmful effect. For example, anticorrosion sprays for cars are used to prevent or delay rusting of iron and steel objects.

The skin's own antioxidant defence protects it from free radical damage. However, when the number of free radicals formed is greater that the capacity of the skin's natural defence system, damage to cells occurs immediately. To help prevent this, antioxidants are added to cosmetics so there is an increase in the natural antioxidant reserves of the skin (see Table 8.1).

Vitamins are the most important type of antioxidants. You can read more about these in Section 8.1.

A single antioxidant exposed to free radicals may become a free radical itself, although a less active one. To help prevent this, antioxidants are used together so there is often a chain reaction that occurs providing better protection.

Antioxidants are a key to age prevention and their daily use in cosmetic products helps reduce UV-induced skin damage.

8.5.2.1 Products Need Protecting Too!. Antioxidants can also prevent other compounds from oxidizing or becoming rancid (in the case of fats and oils) within the cosmetic product itself.

Table 8.1 Some of the most common antioxidants used in cosmetics.

Antioxidant	Source
Carotenoids	Found in yellow and orange vegetables and fruit, such as carrots, butternut squash, yellow and orange bell peppers, pumpkin, corn and sweet potatoes
Lycopene	Gives tomatoes, watermelons and papaya their bold red colour, helps our body make vitamin A
Anthocyanins	Found in berries, currants, grapes, tropical fruits, red to purplish blue-coloured leafy vegetables, grains and roots
Flavonoids	Found in blue and purple vegetables such as aubergine, purple cabbage, purple peppers, purple potatoes and purple onions
Lutein and zeaxanthin	Found in brightly coloured green vegetables such as asparagus, broccoli, Brussels sprouts, celery, peas, spinach and courgettes
Vitamins such as C and E	As you have already read about in Section 8.1

Antioxidants used to protect the product must meet many criteria to be suitable. For example, the ingredient must be effective at preventing product oxidation (at low concentrations), retain its antioxidant activity over a considerable period (so it isn't all used up straight away), be unaffected by heat, light and moisture and, of course, must be safe on the skin. The ideal antioxidant should also be odourless, tasteless and colourless so as not to impact adversely the appeal of the product.

8.6 ANTIMICROBIALS

In recent years, we have often seen reported misconceptions about the safety of long-used traditional preservatives such as parabens (see Chapter 10 for further information), the antimicrobials that were once commonly keeping harmful microorganisms from flourishing in our cosmetics. Also, consumer trends such as natural, 'clean' beauty and vegan suggest a strong desire of consumers to seek ingredients from more natural and ethical sources. This has led to changes in the ingredients commonly used as preservatives and antimicrobials, but what are these ingredients and why are they so important? In this section you'll read about the microorganisms that present a risk to health and safety when they are present in products and learn how preservative systems aim to keep them under control.

8.6.1 The Germs (Microorganisms) Around Us

Now, before you feel the urge to reach for the soap and clean the microbes from your skin, it's important to know that microorganisms are found almost everywhere in our environment, ranging from our skin to some of the most inhospitable places known to humans. Bacteria, a type of microorganism, were one of the first organisms to exist on our planet and most microorganisms live in harmony with their surroundings and with animals and plants that act as their hosts. They may even be performing very important jobs, such as those that live in our gut. When certain types of microorganism end up in the wrong place or change in population, they can become a problem. For example, if *Staphylococcus aureus* finds itself in a wound, a blood infection is possible. When *Cutibacterium acnes* (formerly *Propionibacterium acnes*) increases in population, acne can pop up right when you don't want it!

Cosmetics provide a perfect environment for pathogenic (disease-causing) bacteria to feed and multiply and without preservation we can find ourselves exposed to higher than normal concentrations of the harmful bacteria and other microorganisms. Although this usually may not cause an issue for most healthy individuals, they can present a risk to babies, the elderly and those with a weakened immune system. The main microorganisms that present a risk in our cosmetics include bacteria, yeasts and moulds.

8.6.1.1 Bacteria. Bacteria are one of the smallest and most diverse set of microorganisms found in cosmetics. Many bacteria are just 1 μm in diameter and can reproduce happily at room temperature. To put this small size into perspective, the width of a human hair is 17–181 μm so at least 17 bacteria could fit in the width of a human hair!

Just like us, bacteria are made up of living cells, but a bacterial microorganism is made of one cell only. Each bacterial cell is surrounded by a cell wall that holds everything inside and acts as a barrier between the inside and the outside – rather like human skin. There are two main types of bacteria, Gram positive and Gram negative, which differ mainly in the composition of their cell walls. Gram-positive bacteria have a thick structural layer within their cell wall, whereas Gram-negative bacteria have a thin structural layer, with a protein/lipid layer in addition. Unlike most human cells, bacterial cells do not contain a nucleus (where our DNA is stored) and are known as prokaryotic cells – the DNA is a single thread within the cell. Bacteria differ from each other in their shape and structure, with some types being round and connected together like pearls in a necklace and others being rod shaped or even spiral shaped (see Figures 8.10–8.12). A common

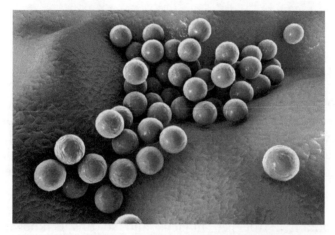

Figure 8.10 *Staphylococcus aureus* (© Kateryna Kon/Shutterstock).

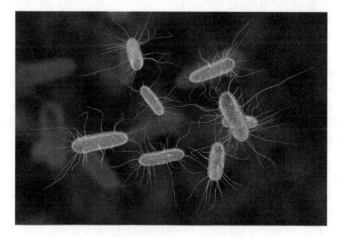

Figure 8.11 *Escherichia coli* (© Kateryna Kon/Shutterstock).

Figure 8.12 *Treponema pallidum* (© Tatiana Shepeleva/Shutterstock).

form of bacteria found in contaminated cosmetics is *Staphylococcus aureus*, a Gram-positive spherical bacterium, known to cause wound infections. Bacteria can be further described as being aerobic or anaerobic, meaning that oxygen is required to live and reproduce or the bacteria can thrive without oxygen.

8.6.1.2 Yeasts. Yeasts are a type of fungi and, unlike bacteria, they contain a true cell nucleus and so this is known as a eukaryotic cell. Yeast cells are egg shaped and they are commonly seen with a 'bud' on the end of the primary cell, which is a new cell growing until it is large enough to break off and become a yeast cell itself. *Candida albicans* is a pathogenic yeast that is commonly found in contaminated cosmetics. *C. albicans* is known to cause athletes foot and infections of the oesophagus, skin and genital area. More commonly, yeasts are found and used in the food industry, *e.g.* in bread making and beer and wine fermentation. *Saccharomyces cerevisiae* (baker's yeast) is illustrated in Figure 8.13.

8.6.1.3 Moulds. These types of microorganisms (made of many cells joined together) can become very large to form what appears like coloured fluff on surfaces where water is usually present. *Aspergillus brasiliensis* (Figure 8.14) is a common form of mould, black in colour and harmful to human

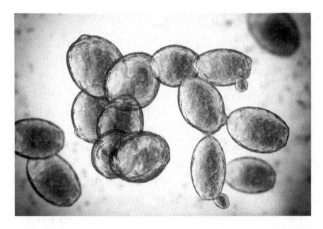

Figure 8.13 *Saccharomyces cerevisiae* (© Kateryna Kon/Shutterstock).

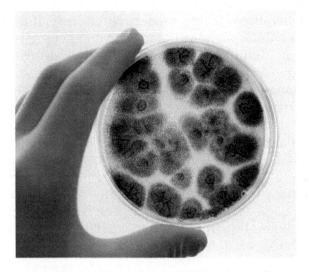

Figure 8.14 *Aspergillus brasiliensis* (© CA-SSIS/Shutterstock).

health. Owing to the large amount of water found in many cosmetics, moulds find it easy to grow.

8.6.1.4 Viruses. Viruses are the smallest of the microorganisms and are made up of just a few molecules of DNA (or RNA) wrapped in a protein membrane (skin). They cannot survive for a reasonable length of time outside living cells, making them a low risk in cosmetics.

8.6.2 Products Need Protecting – 'Preserving'

Preservatives are ingredients added to cosmetic products to protect the product from any microorganisms that get in during storage or use (see Box 8.11). Different types of preservatives act in different ways. Ideally, a preservative will have multiple ways for preventing the growth of microorganisms. Organic acids, such as benzoic acid, transform the environment into one in which the microorganisms cannot grow, simply by reducing the pH of the product. Microorganisms prefer to thrive at a pH of 5–8 in most cases.

Alcohols and phenols, such as phenoxyethanol and parabens, have bactericidal (cell destruction) properties. They prevent the production of new proteins within the cell and, by preventing the production of DNA, prevent the cell from reproducing.

Phenoxyethanol is moderately effective against bacteria, yeasts and moulds, but is restricted to a usage level of 1% in cosmetics in the EU. Because it is only moderately effective, it is often necessary to combine phenoxyethanol with another ingredient that will boost its antimicrobial action. Ethylhexylglycerin is commonly used as a booster.

In addition to ingredients that are added with the sole purpose of protecting against microorganisms, some ingredients that are added for other properties can contribute towards the antimicrobial protection of the product. Pentylene glycol is one example; it is usually added to a cosmetic as a humectant, solubilizer or moisturizer but it additionally acts as an antimicrobial by disturbing the integrity of the bacterial cell membrane, similar to alcohol-based preservatives.

Another innovative type of antimicrobial, based on a microorganism itself, is lactobacillus ferment. *Lactobacillus acidophilus*

BOX 8.11   COMMONLY USED PRESERVATIVES

- Phenoxyethanol
- Benzoic acid
- Butylparaben
- Imidazolidinylurea
- DMDM hydantoin (1,3-dimethylol-5,5-dimethylhydantoin)
- Methylisothiazolinone

bacteria are found in fermented foods such as yoghurt. Unlike different pathogenic microorganisms, this bacterium is known for its probiotic benefits to the body. Although lactobacillus ferment is not a probiotic in the sense that it is a live bacterium, it has good antimicrobial properties. It works in a similar way to organic acids in that it creates an acidic environment and reduces the growth of many pathogenic microorganisms. Generally, mixtures of different antimicrobials ensure the best protection.

8.6.3 What Happens if Cosmetics Are Not Preserved?

First, it should be highlighted that not all products need to be preserved. Products that are naturally hostile towards microorganisms, such as those without water, a low/high pH or a high alcohol content, are unlikely to have microorganisms grow in them. For all other products, preservatives are not only included in cosmetic products to ensure the shelf-life of the product, they also exist to protect from any external factors compromising the integrity of the product. For example, you may dip your finger in a pot of moisturizer, inadvertently contaminating the product with extra bacteria, and the formulation needs to be able to cope with this threat. Buttering toast and then using the same knife to spread jam (Figure 8.15) is a good analogy for this.

Other examples of contamination are in the 'headroom' of a product. This is the space between the product and the lid. Fungal growth can occur if products are left open in areas subject to increased moisture such as bathrooms. Packaging can play an important role here. Think of pump bottles: these can be completely airless in addition to having the benefit of not being 'open' in a wet bathroom (see Figure 8.16). Less preservation may be required in these types of packaging.

8.6.4 How Do Companies Know If Their Products Will Remain Safe If They Become Contaminated?

Challenge testing is a common method used to test for the effectiveness of the preservative against certain types of pathogenic (disease-causing) microorganisms in cosmetics. Challenge testing involves inoculating (deliberating contaminating) a cosmetic product with specific microorganisms and observing their growth

Figure 8.15 The result of using the same knife to spread butter and jam on toast (© gcpics/Shutterstock).

Figure 8.16 Airless packaging (© Isaac Zakar/Shutterstock).

over a specific period of time. If all of the microorganisms decrease in growth by a specified amount before a set time, the cosmetic product is considered effectively preserved.

8.6.5 Antimicrobial Protection on the Skin

Some cosmetic products also have a second function to protect the skin from harmful microorganisms, *e.g.* a soap with an antibacterial function.

The types of ingredients used to protect the skin are similar to those used as preservatives but usually need to be included in higher amounts. Other products that would fall into this category would be anti-dandruff shampoos, foot-care products to reduce foot odour and antibacterial hand products.

In some markets around the world antibacterial and antimicrobial products are regulated as medicines, but in the UK and most of the EU they are usually considered to be cosmetic products. It is important to note that in order to comply with the regulations of a cosmetic product, the antibacterial effect must be a second function and not the primary purpose of the product. For example, an anti-dandruff shampoo primarily cleans the hair but also provides an anti-dandruff benefit, so it is usually considered to be a cosmetic product.

8.7 NATURAL EXTRACTS

The term 'natural' can mean very different things to different people. You will read more about what 'natural' means in Chapter 10, but for now let's think of ingredients that are extracted directly from plants or animals.

Natural herbs and plant extracts have been used by a variety of cultures since Ancient times. The Egyptians are famous for their use of honey and milk to bathe Cleopatra and, for over 5000 years in India, Ayurvedic skincare and medicinal practices have used natural ingredients.

Fast forward to the last 50 years or so and the UK cosmetic industry has evolved its approach to natural extracts and how best to extract these materials for cosmetic use. These natural components were typically created using a solvent, usually water or alcohol, to extract the natural material from its natural source (see Box 8.12). In more recent years, extraction with a solvent has been replaced by the use of supercritical extraction methods, a new direction for natural extraction. The most often used supercritical fluid is carbon dioxide (CO_2). On exposing the natural source to CO_2 at high pressure, the CO_2 disperses into

> **BOX 8.12 EXTRACTION**
>
> Extracting one component from a matrix (collection of components) using a fluid. For example, caffeine is released from tea when the tea is infused in hot water.

the natural source, dissolving the material that needs to be extracted from its starting source, so that it can then be collected. Other extraction methods may also be used. Varying the method can help to obtain different components, flavours or aromas. For example, a lipophilic compound will not extract well, if at all, into a water-based solvent.

For some brands, natural extracts form part of the brand identity and ethos. In 1976, Anita Roddick launched her brand The Body Shop. This was one of the first movements towards creating products based on naturally inspired and ethically sourced ingredients using natural materials from plants rather than animals as the hero ingredients, *e.g.* dewberry and white musk.

Moving on from this, manufacturers started to invest more time and money in researching naturals and the best way to extract their properties. The use of natural extracts can pose a variety of difficulties to anyone formulating products. For example, the way in which a plant is grown and the time at which the material is harvested can completely change the properties of the natural extract (such as how it looks, smells and performs). Also, as natural extracts have gained in popularity, the demand for these materials has risen, which means that manufacturers need to find faster and more eco-friendly/sustainable ways in which to obtain the extracts. For instance, crops that rely on specific weather conditions can be impacted by changes in water, temperature and soil composition. Jojoba is a very good example. Jojoba became a great vegetable alternative to the animal-derived whale oil that used to be used in cosmetics. However, between 2010 and 2013 there was a global shortage of jojoba, causing prices to rise. The amount available was so small that many were forced to use 'Nature-identical' versions (see Chapter 10).

The jojoba oil situation is just one example of how Nature can be unpredictable, but it forced manufacturers to consider how they can continue to provide sustainable natural products. Jojoba is now grown in various locations in both the northern and southern hemispheres so that climate variations in one region can be offset by crops available from another region.

Jojoba oil is sustainably grown but there are other natural plants/fruits that can deeply affect the surrounding environment and wildlife if the material is harvested or uprooted from its natural habitat without care and attention. This is particularly key when considering extracting exotic and rare species. International biodiversity laws (you can look up the Nagoya Protocol to read more) help to protect against the overuse of such extracts in addition to supporting the communities from which they are obtained.

Through natural evolution, some manufacturers have chosen to adopt the 'cradle to cradle' approach (Figure 8.17) when creating natural extracts. This design is about implementing a holistic, economic and social framework around the way in which manufacturers create their natural extracts. They strive to seek systems that are not only efficient but also waste free, for example using all of a particular plant or fruit. The coconut is one example where the whole of the coconut can be put to use. With this approach, six different natural extracts and actives can be created for use in personal care without wasting the fruit.

Another sustainable practice is to make use of something that would otherwise be thrown away. This waste could be from the food or other industrial sectors and then processed to create an ingredient for use in personal care. As this only uses materials

Figure 8.17 The 'cradle to cradle' approach to product development. Reproduced with permission from Active Concepts.

from waste, it also means that no natural products are taken directly from their habitat. As we become more aware of the effects that waste has on our environment, upcycling is becoming more popular.

8.7.1 Producing Natural Extracts Using Stem Cells

Stem cell culture is also becoming more popular. These are cells that have been taken from a living tissue (such as a plant) and placed in a carefully controlled and sterile environment in the laboratory, for example in a Petri dish also containing food for the stem cells and all the essential nutrients needed to encourage growth. These stem cells are then able to reproduce over and over again.

There are many advantages to making natural extracts using stem cell culture, including being able to produce natural materials much faster than *via* the traditional route. A key benefit is also that it limits the impact on the environment and helps to preserve rare or exotic species. Like humans, all stem cells have a life span. However, some cell culturing cells can be 'transformed' into immortal cells and, if kept under the optimum conditions, they will continue to reproduce over and over so there may never again be a need to remove the natural material from its traditional environment. As with most things, this process also has limitations and some ingredients cannot be manipulated in this way, making it not always suitable. Also, the cost of stem cell cultures can often be high, so products containing these extracts may be more expensive to buy.

The trend and growth of naturals show no signs of slowing and manufacturers continue to look at ways in which they can provide safe, effective and ethical natural extracts to keep up with demand.

8.8 DELIVERY SYSTEMS

Delivery systems are methods used to ensure that an active ingredient reaches the area of the body where it is needed. Delivery systems are widely used in cosmetic products and medicines. They are often used to give better penetration of an active ingredient into the skin so that it can work more effectively.

Formulations that contain delivery systems can take many forms, such as gels, creams, adhesive patches, sprays or even instrumental methods employed by specialist skincare clinics. The choice of delivery system will depend on many factors, such as the solubility of the active ingredient and how deeply into the skin the active ingredient needs to penetrate.

8.8.1 Why and Where Are Delivery Systems Used?

How well an active cosmetic ingredient works is generally helped if it can be delivered directly to the site of action in the epidermis (upper layer of the skin) but, as you've read throughout this book, penetration through the skin is quite difficult.

If an ingredient is able to pass into the skin, the delivery of an active ingredient needs to be carefully controlled to make sure that it doesn't enter the bloodstream. Selection of the appropriate delivery system allows for an ingredient to be released gradually over time or for it to be delivered to a specific part of the skin. This will help to ensure that the ingredient works better to do what it is designed to do. For example, although these delivery systems cannot be seen with the naked eye, they can be manufactured in different sizes, allowing some to reach lower levels of the epidermis than others.

You might expect that delivery systems would be restricted to expensive cosmetics, but this is not true. Delivery systems are also used in relatively low-cost products such as 'body-responsive' antiperspirants and deodorants. Microcapsule shells made of a wax-like material and containing antiperspirant salts, deodorants or antimicrobial ingredients are designed to break apart in the humid and warm environment under the arms, resulting in the release of these ingredient for a prolonged effect.

Delivery systems can also be used to protect sensitive active ingredients. As we saw earlier, retinol (a type of vitamin A) can be easily destroyed if exposed to the air and can react with a number of other ingredients typically used in cosmetic formulations. However, if retinol is placed in a suitable delivery vehicle such as a microcapsule, then it can be protected until the product is applied to the skin.

8.8.2 Examples of Delivery Systems

8.8.2.1 Time-controlled Release. Time-controlled release allows for a steady and consistent amount of the 'active' to be delivered into the skin. An everyday example you may be familiar with is the adhesive transdermal patch used for some medicines such as nicotine patches.

Time-controlled release technology has also been used in the cosmetics industry. One example involves trapping the ingredients inside a polymer that is added to a cosmetic formulation so that the ingredient can break out over time once it touches the skin. This technology has been successfully used in lipsticks, pressed powders and similar anhydrous (do not contain water) colour cosmetics. Vitamin complexes, hyaluronic acid and salicylic acid are examples of cosmetic ingredients that are delivered using this technology.

Microencapsulation also offers time-controlled release of actives, for example the 'body-responsive' antiperspirant-containing microcapsules described in the previous section.

8.8.2.2 Physical Barrier Disruption. The upper layer of our skin forms a waterproof barrier to protect us from the outside world and to keep water inside. You will have read more about this in Chapter 5 and that it is made up of a lipid structure.

So that active ingredients, especially water-soluble ingredients, can be delivered to the skin, some temporary disruption of this lipid structure may be necessary. One approach used successfully for many years is the liposome. Liposomes are made up of lipid (phospholipid) bilayers that form an enclosed structure.

Because the phospholipids in the liposomes are similar to the lipids found in the stratum corneum and the very small size of the liposome (typically 100–500 nm), the body allows liposomes through the stratum corneum to deliver the encapsulated materials to the desired location (see Figure 8.18).

There are many cosmetic ingredients that can be successfully used in liposomes, including vitamins, natural extracts and antioxidants.

In addition to liposomes, there are even smaller nano-sized structures available (typically 10–40 nm). The lipids used in the structure of these nano-vesicles mimic the structure and carrying ability of natural lipoproteins found in the skin. These tiny

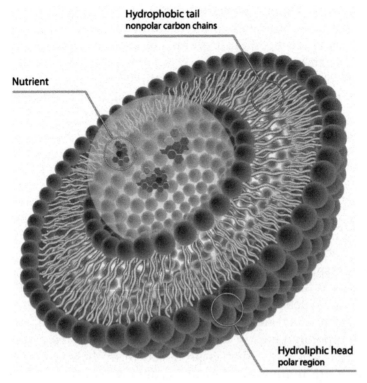

Figure 8.18 An example cross-section of a typical delivery system used in cosmetic products (© matike/Shutterstock).

structures can entrap a variety of active ingredients and accumulate in the upper skin layer (stratum corneum), then degrade harmlessly to provide sustained release of the active ingredients into lower skin layers.

8.9 ANTIPERSPIRANT AND DEODORANT EFFECTS

8.9.1 What Is the Difference Between an Antiperspirant and a Deodorant?

Let's start by looking at the differences between the two different products, as they can often be confused with each other.

Deodorants are products that are designed to cover or mask body odour without affecting the amount of sweat produced. Sweat itself doesn't actually have an odour, the odour is caused by the interaction of sweat with our skin microbiota, the bacteria on the skin. Various bacteria on the skin feed on the sweat, digesting the fats and proteins and changing the sweat composition all together. It is this change that can sometimes result in bad body odour.

Traditional deodorants contain antimicrobial agents that will reduce the number of bacteria on the skin and inhibit the change in odour. Ingredients such as triethyl citrate also help to reduce body odour as various bacteria attack the triethyl citrate. By attacking this material, it actually changes it into citric acid. Citric acid in turn changes the acidity (pH) of the underarm area so that it becomes an environment in which bacteria cannot grow and thrive, eliminating the ability to feed off of the sweat and cause the odour. In the future we may see deodorants containing materials that work *with* the bacteria on our skin to prevent body odour.

Antiperspirants, on the other hand, block the sweat glands and prevent them from producing sweat in the first place. These types of products contain active ingredients such as aluminium chlorohydrate or aluminium zirconium chlorohydrate, which form temporary plugs in the upper parts of the pores, reducing the amount of sweat that is produced from the sweat glands. As these are temporary, they are generally removed over time by the skin's natural renewal process or through washing. Antiperspirants have a natural deodorizing benefit as they limit the amount of sweat produced, so reducing the negative interaction between the skin bacteria and sweat.

8.9.2 Why Do Antiperspirants and Deodorants Come in Different Formats?

The format of these products can vary greatly. Common formats are aerosols, pump sprays, roll-ons and solid sticks. Antiperspirants and deodorants come in these many formats to meet consumer preferences rather than for functional purposes. For instance, aerosol products provide a quick drying time and can also be used by the whole family without compromising on hygiene, whereas roll-ons are slow to dry but provide very good coverage and effectiveness. The latter is also true of solid sticks, but they can leave a white residue on both the skin and clothing. It is very much down to personal choice which is best for you!

8.10 CONCLUSION

In our look at the science behind active ingredients, we've considered why certain types of ingredients are important both to the integrity of the products and to delivering product benefits that are vital to maintaining healthy skin, nails, hair and teeth. We learnt how vitamins are not just something we eat, how UV filters don't just protect humans from the Sun's rays but also can be used to protect the product, how antioxidants can help protect our cells from attack, that the bacteria around us can ruin our products (in addition to making them unsafe to use) and how stem cells can be used to produce Nature-identical substances without impacting the environment or posing any risk to biodiversity. We've looked at a wide variety of ingredients, but the personal care industry and the science on which it relies are ever evolving, with companies constantly innovating and using new ingredients and technologies in order to create more and more efficacious and novel products. Who knows, in the years to come perhaps this chapter will contain an entirely different set of important ingredients!

CHAPTER 9

Testing and More Testing – The Science Behind Keeping Your Skin Safe and Healthy

STEPHEN KIRK

SK-CRS Ltd, UK
Email: stephen.kirk@sk-crs.com

9.1 COSMETIC PRODUCTS – HOW WE KEEP YOU AND YOUR SKIN SAFE AND HEALTHY

Before a new cosmetic or skincare product can be launched for sale, the mixture in its container (the formulation) has to be tested and checked in a number of different ways to ensure that the consumers who use it will not come to any harm. This process, which is often very complex and time consuming, involves checking a large number of different physical and chemical properties of the formulation.

Checks are also carried out to ensure that any claims made about the product, for example, the sun protection factor (SPF) for a suncream or the amount of moisturization for a skincare

---

Discovering Cosmetic Science
Edited by Stephen Barton, Allan Eastham, Amanda Isom,
Denise McLaverty and Yi Ling Soong
© The Royal Society of Chemistry 2021
Published by the Royal Society of Chemistry, www.rsc.org

product, are fair, honest and truthful. Indeed, these checks are part of the covenant between brand owners and their customers to ensure that products do exactly what they say they're going to do. There is also a legal requirement for these checks to be made.

In addition, the safety of every chemical ingredient in the formulation has to be checked by a suitably qualified expert to ensure that they will not cause any harm to the consumers using the product. This phase in the development and launch of a new product is usually known by the term *safety* or *risk assessment*.

Although it may seem that the launch of the product marks the end of the process, brand owners will keep each product that they sell under their watchful eye. This process, which is often known as *post-market surveillance* (or *cosmetovigilance*) is designed to enable brand owners to spot the products that may be causing problems for those who are using them.

These issues may be related to the skin tolerance (acceptance) of the product or more fundamental aspects such as failures of the packaging in which the product is sold. Continuous monitoring will ensure that any such issues, whether trivial or serious, are identified in a timely manner so that appropriate measures can be taken to correct the cause if required.

9.2 STABILITY TESTING – MAKING SURE A PRODUCT IS FIT FOR PURPOSE

During the development of a cosmetic product, the formulation scientist who is working on the product will use their expertise to identify a number of fundamental microbiological and chemical attributes that will be used to set the quality standards for the product. Uniform, often global, testing methods will then be used to measure each of these characteristics against the specific standards that have been set for each. Collectively, these testing methods and standards are referred to as *stability testing*.

Compliance with these standards will ensure that the consumer can use the product in the knowledge that they will not come to any harm during the time that it is being used. This is the principle that underpins stability testing of a new cosmetic product.

Figure 9.1 How not to store your shower or bath products.

9.2.1 Microbiological Testing – Will It Go Mouldy?

Because of the way in which cosmetic products are made and then used and stored by consumers, they can be at risk of becoming contaminated by microorganisms such as bacteria and moulds. For example, products such as shower gels, shampoos and hair conditioners are often kept on the base of shower enclosures (Figure 9.1). The result is that they are regularly exposed to both water from the shower head and also the rinse water from the person using the shower. This continual drenching gives many opportunities for water containing microorganisms to enter a product's container.

As a result, strict criteria have been developed by both cosmetic companies and government regulators to prevent products from becoming infected by microorganisms through their normal use by consumers.

The first phase in ensuring the microbiological quality of a cosmetic product is concerned with the integrity of the chemicals and mixtures that are used to make the product – the 'cosmetic ingredients'. The source and nature of an ingredient are the key characteristics used in the decision-making process of setting a microbiological standard for that ingredient.

Ingredients that are made from natural sources will be more susceptible to infection by microbes if care is not taken during their processing. For example, soil residues can often pollute plant materials during harvesting, transporting and processing (Figure 9.2).

Figure 9.2 Soya plant seeds destined to be made into cosmetic ingredients (Credit: John Lambeth, Pexels https://www.pexels.com/photo/orange-beans-dropping-from-machine-2965711/).

Subsequently, if the plant materials are not adequately cleaned or washed prior to final processing, there is a risk that any microbes present will be carried into the ingredient itself.

Other classes of ingredients, for example those which are predominantly nutritional sources for microorganisms, will also be prone to infection if not handled correctly both before and during the manufacturing process.

The source of water used to make many cosmetics can also be a source of microorganisms if its purity is not closely controlled. Domestic tap water is usually not pure enough to be used to make cosmetics.

It is quite common for ingredient manufacturers to include chemicals known as antimicrobial preservatives in their materials to prevent microbial spoilage before the ingredient is added into a finished product. This process will prolong the shelf-life of the cosmetic ingredient.

Natural ingredients such as mineral-based powders, for example talcum powder or those based on other natural powders, may also be susceptible to microbial growth. In the case of these materials, the mine from where they are sourced is carefully inspected and material tested to ensure that any minerals taken from the mine are free from microorganisms. As a further safeguard, these powders are often treated with ultraviolet (UV) light or chemicals such as ethylene oxide that are harmful to microbes and will kill any microbes that may have infected the materials during processing.

However, there are many cosmetic ingredients available that will not support the growth of microbes. Synthetic (manufactured) materials, particularly if they are supplied in a powder format, do not contain water and can be counted amongst such materials. Similarly, the coloured ingredients used to manufacture make-up products are often powdered and as such will generally not support the growth of microorganisms (Figure 9.3). Nevertheless, these materials will always be allocated a microbiological specification and associated checking as a precautionary measure.

Once they are satisfied that the microbiological quality of their ingredients is satisfactory, formulation scientists must then turn their attention to the microbiological standards of the formulations on which they are working. This is particularly important for products that are going to be used by certain susceptible groups of consumers. These consumers include those at the extremes of age such as infants and the elderly. Individuals with

Figure 9.3 Cosmetic colorants in powder format (Credit: Thom Masat, Unsplash https://unsplash.com/photos/8kLFdQh6scA).

> **BOX 9.1 DID YOU KNOW?**
>
> One of the best ways for consumers to help protect against microbial contamination is to stick to the following advice:
>
> - Don't share cosmetics with anyone. You may be sharing germs!
> - Don't add water or saliva to cosmetics such as mascara. You may be adding bacteria or other microorganisms and you'll also be watering down a preservative that's intended to prevent microorganisms from growing!

a compromised immune system such as those undergoing certain medical treatments would also be counted among this group of consumers.

Use on body areas that are readily susceptible to infection will also be a major point in determining the microbiological quality standards of formulations. Cosmetics that are applied to so-called mucous membrane surfaces (for example the eyes and lips) pose a hazard for potential microbial infection. Similarly, use on damaged or diseased skin, for example open wounds or eczema, could also lead to infections in the affected skin. See Box 9.1.

Opportune microbial spoilage of cosmetics is generally held in check by the use of the same antimicrobial preservatives as described earlier. This is a particular risk factor for products that are mostly made of water. It is also true for cosmetics that contain substantial amounts of ingredients that can act as nutritional sources for microorganisms. This type of ingredient includes those made from starch as a starting point or other carbohydrates.

In contrast, formulations that contain significant amounts of organic solvents such as alcohol or those that are inherently highly acidic or alkaline are at less risk of infection. However, it is important to note that the addition of preservatives to formulations should not be used to overcome microbial contamination that may arise during the manufacturing process. It is for this reason that strict microbiological standards are also applied to the processes used to manufacture the finished product. These processes are described in industry good manufacturing practice (GMP) guidelines, which are designed to ensure the consistent quality and safety of the finished goods.

Although many chemical substances can have preservative properties, the cosmetic regulations of many countries will allow only certain chemicals to be used in cosmetic products.

Once the formulation scientist has decided which preservative (or preservatives) they are going to use, they will then test the formulation to check if the chosen chemical will preserve their formulation in the event that it becomes infected by microbes. This is done by challenging the preservative-containing product by deliberately adding microorganisms and measuring whether they continue to multiply. This test is commonly known as the *preservative efficacy test* (PET).

The objective of the PET is to assess the effectiveness of the preservative system following deliberate infection with a range of different microorganisms that are representative of those which may cause infectious human diseases. The PET generally conforms to well-established methods described in international guidelines. However, it does not have to be conducted on every type of product. Like the ingredients described earlier which are not prone to microbial infection, products that contain high levels of alcohol (for example perfumes) will not be susceptible to infection. Therefore, in such cases, there would be no need to include a preservative or to conduct a PET.

Similarly, products sold to the consumer in airtight, sealed containers would not be at risk of microbial spoilage. This type of packaging format would prevent the consumer from coming into direct contact with the formulation within the packaging. Again, there would be no need to conduct a PET for these formulations. The same would apply to single-use products such as certain luxury skincare serums that are sold in vials or ampoules and contain only a sufficient quantity of the formulation for a single application to the skin.

However, sometimes things do not always go to plan and microbially contaminated products are removed from retailers' shelves as a result of enforcement action by the authorities who ensure the safety of marketed cosmetics.

Important information about cosmetic products that pose a risk to the health and safety of consumers due to bacterial contamination and that are withdrawn from sale in the European Union (EU) is conveyed by *The Safety Gate Rapid Alert System*, which permits a rapid exchange of information between EU/EEA member states and the European Commission about dangerous non-food products.

9.2.2 Chemical Stability – Will My Product Change Colour?

There are several chemical and physical features of a cosmetic that are important in ensuring that a product stays fit for purpose while it is being used by a consumer. These include its thickness (viscosity), pH, colour and smell. These aspects all play a role in how the consumer perceives the product and its ease of use. Ultimately, these characteristics will often determine whether a consumer buys the product again.

With this in mind, cosmetic companies have developed a number of experimental methods to measure these features against preset standard values, collectively known as *stability testing*. The results of this testing are then used to set a 'shelf-life' or 'durability of use' for the product.

Like the microbiological preservative efficacy test, the physico-chemical parameters of the formulation are also measured over time but with different stresses of extremes of temperature (very cold to very hot, usually in cycles) and light. These conditions are designed to mimic (in an accelerated way) the typical environmental conditions that the product may experience during its lifetime. This includes storage after manufacture (both before and during transport to the shop where it will be sold), storage on the shelving in shops and then finally storage by the consumer in their home.

Although the stability testing is conducted out of sight of the consumer, the parameters that are measured are important in determining whether or not a consumer will buy the product again. For example, many products contain a perfume that is designed to give the product added consumer appeal. Indeed, consumers very often buy a product based on how it smells rather than for its benefits to their skin.

With this in mind, high temperatures can often have a detrimental effect on perfumes that are included in products. Such temperatures can cause a perfume to lose some of its characteristic smell (odour notes). Similarly, some products may develop 'off-smells' during storage, for example rancid smells in formulations that contain high levels of certain fats and oils. If these situations occur, the consumer will be deterred from buying the same product again in the future.

Testing for colour fastness is another key stability attribute for cosmetics that are coloured in some way or another. This testing is usually performed by exposing the product to intense artificial sunlight for predetermined periods of time. The objective of this test is to check whether the sunlight has any detrimental effects on the colour of a formulation should the product be stored in direct sunlight on a window ledge or window-facing vanity unit in the home. In addition, colour fastness during storage in a shop window or on a shop merchandising unit is also checked by exposing products to the same type of lighting conditions as commonly found in retail stores.

If any such detrimental effects are discovered, corrective action can be taken to prevent this happening during use of the product by the consumer. Such action could be achieved by the inclusion of a UV filter in the product; UV light is most often the part of sunlight that causes products to change colour. The detrimental effects of UV and visible light on products can also be prevented by the use of intensely coloured packs or by the inclusion of UV filters in the plastic materials used to manufacture a product's packaging.

Although exposure to sunlight is the principal reason why some cosmetics unexpectedly change colour, some of the actual ingredients used can also have the same effect. Vanilla-based chemicals, which are often used in sweet-smelling fragrances, are known to develop yellow discolouration as the product that contains the substance ages. Stability testing is designed to detect such changes before the product is sold to the consumer. If a vanilla-containing fragrance is found to turn yellow during storage, the formulation scientist will have no option but to modify the fragrance to remove or replace the substance responsible for the colour change.

Assessing the thickness (or viscosity) of a cosmetic is also a very important part of a stability testing programme, particularly for products that are sold to the consumer in unusual bottles

Figure 9.4 Common 'tottle' bottles (© Shutterstock).

such as those which are stored on their tops. These bottles are commonly known as 'tottles', as illustrated in Figure 9.4. Liquid products that are sold in 'tottles', for example hair conditioners, suncare and baby skincare products, are designed with a specific thickness in mind to prevent them from leaking from their containers during storage. Their viscosity is also set at such a point as to ensure that they do not 'flood' or 'dribble' from the bottle during use by a customer.

Stability testing at the extremes of cold and heat over prolonged periods of time is designed to show if a product is liable to thin during storage, in other words, if its viscosity will decrease. Should this happen, the product will more easily leak from a bottle, which could result in damage to a consumer's property. Conversely, the viscosity of some products may increase, making it difficult for the customer to dispense the product from the bottle.

9.2.3 Period After Opening (PAO) and Shelf-life – How Long Will It Be Okay to Use?

Once the stability testing programme is completed, the formulation scientist will use the information generated, often primarily the results of the PET, to determine the formulation's 'durability of use'. This term is used to express a time scale during which the cosmetic will be safe and efficacious to use after it was first used.

In the UK and Europe, where the stability testing results indicate that the durability of use is at least 30 months, a date known as the 'period after opening' will be set for the product. This date, which is usually expressed in months, is indicated by a standardized symbol, shown in Figure 9.5. The date in months is shown as a number within the body of the jar symbol. Conversely, cosmetics for which the durability of use is less than 30 months (UK and EU) will be labelled with a 'best before' date instead.

However, showing a period after opening symbol is not required for all cosmetics. For example, such a date would not be applicable for so-called 'single-use' products. As the name suggests, these are products that are sold in small quantities, either as a liquid or a solid, which give sufficient formulation for only a single application by a consumer.

Similarly, products that are not susceptible to microbial contamination are exempt from having to have a period after opening symbol on their label. These products include those which are made predominantly of inert substances such as ethyl

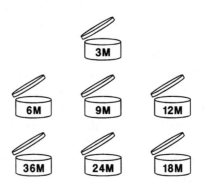

Figure 9.5 'Period after opening' symbol (© Shutterstock).

alcohol. Products that are sold in pressurized containers such as hairsprays and deodorants would not be susceptible to microbial spoilage. This is because their packaging is a closed system, which prevents the consumer from introducing microorganisms into the product.

However, despite the extensive stability testing that is conducted with cosmetic formulations, on isolated occasions failures can occur. On rare occasions, these failures can represent a hazard to the consumer and it is partly for this reason that cosmetic companies operate complex systems to monitor the performance of their products once they are launched. These systems are commonly known collectively as 'post-market surveillance', as discussed in more detail in Section 9.6.

9.3 SAFETY ASSESSMENT OF COSMETIC INGREDIENTS AND FINISHED PRODUCTS – IS MY CHOSEN PRODUCT GOING TO HARM ME?

9.3.1 Hazard and Risk – The Tale of a Shark and the Swimmer

It is a fundamental requirement that all cosmetic products are safe, which means that they must not cause any harm to consumers, whatever their age, gender or ethnic origin.

In many countries of the world, brand owners are obliged by local laws to ensure that the safety of the ingredients in their products and the finished products themselves will not cause any harmful effects to consumers using the products. This is achieved through a structured review process that involves a detailed check of the product and its ingredients by a qualified expert in cosmetic safety. The process itself is often known as a 'risk' or 'safety assessment'.

The EU led the way in developing this philosophy as it is applied to cosmetic products. The EU model was first introduced in 1976 and since then many other countries have adopted the same model as their own. Although the EU model has been in existence for more than 40 years, risk assessment processes have not stood still. Accordingly, the EU legislation, and many others for which it is a mirror, have also been updated to keep abreast with these new methods and philosophies.

The first step in the risk or safety assessment process entails a detailed check by a qualified expert (safety assessor) on whether the chosen ingredients are likely to cause any harm to consumers. This is commonly known as the toxicity or hazard of the ingredient or substance. In Figure 9.6, it is the shark that represents the hazard of the situation.

Whether or not the hazard becomes a real threat is very dependent on a number of key factors that include how and where a consumer is exposed to the hazard itself. The amount (or concentration) of each ingredient or substance in a cosmetic also plays a key role in their potential to cause harm to the consumer. In other words, the concentration of a substance determines whether the hazard of the substance becomes visible. This is known as the consumer's *exposure* to the substances.

The safety assessment process is designed to quantify the likelihood (or *risk*) of the hazard becoming visible in consumers

Figure 9.6 The tale of the shark and the swimmer (Credit: Lubo Minar, Unsplash; https://unsplash.com/photos/ECxwQjLRwLA).

while they are using a product in the normal way that it was designed to be used. This is expressed as the following ratio:

$$\text{risk} = \text{hazard} \times \text{exposure}$$

In 'the tale of the shark and the swimmer' (Figure 9.6), a shark has been seen swimming in the sea (*hazard*). However, the swimmer can avoid the shark by not entering the sea (*exposure*), thus preventing injury and minimizing *risk* from the shark.

With this concept in mind, there are several different factors that may contribute to a consumer's exposure to each of the substances in their cosmetics (see Box 9.2).

Once the safety assessor is satisfied that they know all they can about the hazard of the ingredients in a product and how a consumer is likely to use it, they can then complete the risk assessment process for the finished product.

At this point, it is important to note that 'risk–benefit' considerations, as applied to medicinal products (where side effects are occasionally seen), do not apply to cosmetics. In effect, the use of cosmetics by consumers has to be 'risk free'. In other

---

BOX 9.2 DID YOU KNOW?

Factors that affect a consumer's exposure to the ingredients in their products are as follows:

- The quantity of each substance in a product.
- Where and when the product is applied.
- The number of times the product is used – multiple times in the day, weekly or monthly.
- How much product is used on each occasion – more is not always better!
- Whether the product stays on the skin – skincare moisturizer.
- Whether the product is rinsed off the skin – shower gel.
- Those 'in-betweeners' – hair conditioners that are left on for a while and then rinsed away.
- Whether the product can be swallowed or breathed in – lipstick or hairspray.

words, the product cannot be sold if the safety assessor decides that it poses a risk to consumers.

9.3.2 Identifying the Hazard Characteristics of a Cosmetic Ingredient

Before the safety assessment process can be started, the safety assessor needs to find the toxicity data that relate to each of the substances in the formulation. Sometimes the search will also take into consideration mixtures of substances. This information can be found in a number of diverse locations, both as hard-copy printed documents and electronically in databases accessed through the Internet.

These information sources will reveal the outcome of historical safety testing that has been conducted with each substance or blend of substances.

Hard-copy information is usually data that have been collected for occupational handling purposes (to assure worker safety) and is presented in a regulatory document known as the *safety data sheet* (SDS). However, the SDS information is often limited and basic in nature, which then requires the safety assessor to seek more detailed evidence from other scientifically sound and robust sources. These other sources could include scientific reviews conducted and reported by expert bodies such as the US Cosmetic Ingredient Review, the EU Scientific Committee on Consumer Safety or the Joint FAO/WHO Expert Committee on Food Additives.

Conventionally, toxicity data for substances used as cosmetic ingredients have been produced in experiments that required the use of laboratory animals. However, increasingly, this type of testing is viewed by regulators and consumers alike to be unethical. In addition, some Non-Governmental Organizations (NGOs) also present certain reasoned arguments why testing in laboratory animals is invalid and therefore could not be used to justify the human safety of ingredients tested in that way.

Consequently, to recognize these concerns, the use of laboratory animals for the safety testing of cosmetic ingredients was banned in the EU in 2009; testing of cosmetic products was banned in the EU in 2004. As a result, it has been necessary for the suppliers of cosmetic ingredients for the European market to use alternative, non-animal testing methods to prove that their new ingredients will not cause any harm to consumers, in other words, they are safe to use in the manufacture of cosmetics. However, at the time of writing, some countries outside the EU still require the use of animal testing data to justify the safety of cosmetic ingredients. Indeed, in some countries that require cosmetics to be registered before sale, animal testing data for the product itself are also required as a condition of sale in those countries. You can read more on this in Chapter 10.

Nevertheless, rapid progress is being made in these countries aimed at modernizing their cosmetic regulatory framework to remove the obligation for animal testing of ingredients and finished products.

9.3.3 Testing New Cosmetic Ingredients Using Non-animal Alternative Methods

Although animal testing of ingredients is no longer allowed in Europe, cosmetic safety assessors still require certain toxicity data if they are to justify the safety of new substances as cosmetic ingredients. A number of new testing methods have been developed by both government-sponsored research laboratories and the cosmetic industry itself, both regionally and globally. One such government organization is the EU Reference Laboratory for Alternatives to Animal Testing or EURL ECVAM. Similarly, Cosmetics Europe, the European Cosmetic Industry Association, has developed an industry-led strategy to develop new, alternative methods to animal testing.

This type of safety testing of chemical substances is called *in vitro* testing or testing using non-animal methods. The foundations that underpin this philosophy were first described in the 1950s by two British scientists, William Russell and Rex Burch, in their influential textbook entitled *The Principles of Humane Experimental Technique* (see Further Reading). These guiding principles are often known as 'The 3Rs' or the 'Replacement, Reduction and Refinement' of methods concerned with the use of laboratory animals.

Although not all 'end points' concerned with the safety of cosmetic ingredients can be studied using *in vitro* techniques, officially endorsed methods have been developed to assess the most important ones. These include techniques to measure the potential for individual substances to cause irritation or allergic reactions in the skin of consumers, and also methods to determine if substances can be injurious to the eyes.

In addition, other methods have been developed that allow research scientists to study whether cosmetic ingredients possess the ability to damage the genetic material of consumers' skin and other organs; the latter is known as 'mutagenicity' and 'genotoxicity' testing.

At their extreme, substances that are found to possess the ability to damage genetic material can often lead to the development of cancerous changes in the skin and other organs of the body. Hence their use as cosmetic ingredients is very closely controlled, which can often lead to them being banned as cosmetic ingredients.

In vitro testing is performed using cell culture techniques, either as layers of cells (monolayers) or more recently clusters of cells that resemble microscopic organs of the body such as the liver or lungs. This is commonly known as tissue culture.

In many cases, the cells that are used to form the monolayers are legacy materials, being derived from tissues and organs of laboratory animals. However, increasingly human tissue samples are being used as sources of cells for *in vitro* testing.

In vitro testing generally follows a set process, which for simplicity is shown in Table 9.1.

9.4 PRODUCT CLAIMS – WILL MY PRODUCT DO WHAT IT SAYS ON THE TIN?

As explained earlier, it is a fundamental requirement that cosmetic products must be safe for use by all consumers, irrespective of their age or gender. However, there are also certain guiding principles that companies should adhere to ensure that consumers will benefit from the claims made for a product.

9.4.1 Sun Protection Testing – Will My Cream Protect Me from the Harmful Effects of the Sun?

A very important group of cosmetics are those which protect the consumer's skin from the well-known damaging effects of the Sun. These products can range from traditional suncreams (or

Table 9.1 The *in vitro* testing process.

Process step	Description
Step 1	Reconstitute some frozen cells into an appropriate nutritional cell culture fluid and place in the wells of a tissue culture plate (Figure 9.7)
Step 2	Incubate the reconstituted cells in the tissue culture plates until the cells form a complete layer (monolayer) on the bottom of the tissue culture plat4e (Figure 9.8)
Step 3	Dilute the ingredient or formulation to be tested in nutritional cell culture fluid (Figure 9.9)
Step 4	Incubate the tissue culture plate containing the cell monolayer and test the ingredient or formulation for the required period of time
Step 5	Remove the tissue culture plate from the incubator, remove the nutritional cell culture fluid and wash the monolayers carefully a number of times
Step 6	Replace the nutritional cell culture medium with a new medium containing a special chemical dye that is designed to be taken up by cells of the monolayer which have not been damaged by the test ingredient or formulation and re-incubate for a period of time (Figure 9.10)
Step 7	Repeat the rinsing process and then inspect the tissue culture plate under a microscope for evidence of damage to the monolayers. No staining in the cells is evidence of cell damage
Step 8	Review the results obtained and analyse them statistically to allow conclusions to be drawn about the potential for the ingredient or formulation to cause injurious effects in consumers

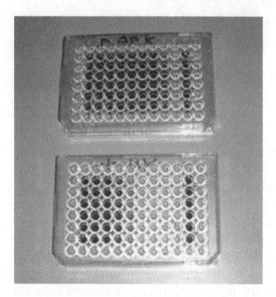

Figure 9.7 Inspecting the cell monolayer for evidence that the cells have been damaged by the test ingredient or formulation. Image courtesy of XCellR8, Daresbury, Cheshire, UK.

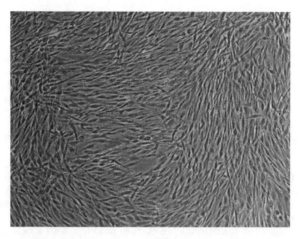

Figure 9.8 A monolayer of human skin cells used in the safety testing of cosmetic ingredients and formulations (Image courtesy of XCellR8, Daresbury, Cheshire, UK).

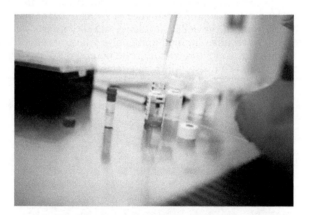

Figure 9.9 Diluting an ingredient or formulation to be tested in nutritional cell culture fluid (Image courtesy of XCellR8, Daresbury, Cheshire, UK).

beach products) for sun protection during outdoor activities, to skin moisturizers that are used every day. Cosmetic companies now recommend the regular use of sun-protecting skincare products to prevent the signs of ageing in exposed areas.

In brief, sunlight is composed of light of different wavelengths, of which UV light is considered to be the most damaging to the skin. UV light is further subdivided into three different groups: UVA, UVB and UVC. UVC light is mostly prevented from

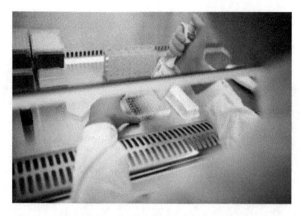

Figure 9.10 Applying the diluted ingredient or formulation to the cell monolayers in a tissue culture plate (Image courtesy of XCellR8, Daresbury, Cheshire, UK).

BOX 9.3 DID YOU KNOW?

Because of characteristic differences in their wavelengths, UVA and UVB rays can penetrate to different depths of the skin.

---

reaching the surface of the Earth by the atmosphere; UVA and UVB rays are able to pass through the Earth's atmosphere to then land on the skin of consumers (see Box 9.3).

As we can see in Figure 9.11, UVA rays are able to reach the deeper layers of the skin. In fact, they are able pass through the whole depth of the skin to reach the fatty layer (subcutis), which sits immediately beneath the skin itself. As the rays pass through the different layers of the skin, they can cause damage to the various tissues in the skin that give it its characteristic feel and structure. For example, one such effect is to cause the breakdown of collagen, which is an important constituent that contributes to the skin's rigidity and structure. The visual effects of this damage are the classic signs of skin ageing, wrinkling probably being the most striking. It is for this reason that this type of UV light is known colloquially as UV Ageing.

UVA light is also responsible for the production of melanin. Melanin is the biological colour that gives the skin its brown or darker shades in the epidermis, the upper layers of the living skin. It is produced in specialized skin cells called melanocytes,

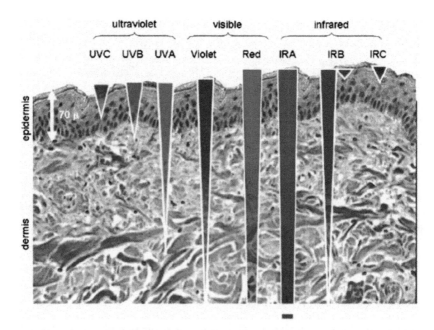

Figure 9.11 How deep will the Sun's rays penetrate? (Source: EC Scientific Committee on Emerging and Newly Identified Health Risks, Health Effect of Artificial Light 2012).

which are usually found scattered randomly throughout the epidermis. It is the role of melanin in the skin to act as a natural sun filter to protect it from UV damage.

In contrast, UVB rays can penetrate through only a few of the outer layers of the skin. It is this UV light that is responsible for the burning effects of sunlight known as UV burning.

Although the resulting sunburn can be quite uncomfortable but short term for the consumer, both UVA and UVB light have a more important and potentially dangerous effect on the skin of consumers. Specifically, it can give rise to malignant cancers in the skin of some individuals, which can be fatal if not detected and treated at an early stage. Consequently, because sunlight exposure can produce these potentially dangerous effects in the skin of consumers, products that claim to protect the skin from these effects must be rigorously tested before they can be sold to consumers.

Traditionally, this testing is conducted in two phases, first to assess a product's ability to protect consumers from burning and second to assess how effective the product is at protecting the skin from the ageing effects due to UVA light. The results of this testing are commonly recognized by two specific terms and symbols found on these types of products, namely the sun protection factor (SPF) and UVA star rating.

The UVA star rating system is also a mathematical concept but one that does not require any specific testing to be achieved. It is an expression of the amount of UVA radiation absorbed by a product compared with the amount of UVB radiation absorbed by the same product.

Figure 9.12 illustrates the UVA star rating symbols that can often be found on sun protection products sold in the UK.

Historically, testing to assess the SPF rating of suncare products has been performed by human volunteers. In the first phase of this testing, artificial sunlight of standard wavelength (generated by a machine called a solar simulator) is shone onto the backs of volunteers and the shortest time taken to cause the volunteer's skin to burn is measured; this time is known as the minimum erythemal dose (MED). Figure 9.13 illustrates a typical solar simulator used for measuring SPFs.

Once the first phase of the testing is completed and the skin of the volunteers has recovered from the sunburn, the test is repeated. However, in this second phase, a standard amount of the product to be tested is applied to marked areas of skin on the back of each volunteer before the second exposure to UV light. Once again, the time taken for a slight sunburn to be detected is measured.

If the product has been effective at protecting the volunteer's skin from burning, the time taken for their skin to show signs of burning should have been extended. A mathematical calculation is then performed that gives the ratio between the two values. This ratio is the figure used to express the SPF rating of the sun product. For example, an SPF 30 product will protect the skin

Figure 9.12 One common UVA star rating system (Source: Walgreen Boots Alliance).

Figure 9.13 Artificial sunlight generator to measure SPFs (Courtesy of Solar Light, Glenside, PA, USA).

from burning for a period 30 times longer than the time that would be taken for unprotected skin to burn.

Although this method has been used for many years (since 1974), it is increasingly being recognized that it is putting the many volunteers who participate in the testing at risk of the skin-damaging effects of sunlight. Consequently, the cosmetic industry in Europe in particular has been developing a new, more ethical testing method that does not use human volunteers. This method will be known as '*in vitro* SPF testing' once it has been approved by the various global bodies who set standards for scientific testing.

9.5 SAFETY-RELATED CLAIMS MADE ON SOME PRODUCTS

9.5.1 Will Your Product Damage My Eyes or Make Them Sting?

Historically, in most countries, including Europe, the testing of cosmetics to assess if they might damage the eyes of consumers was performed on laboratory animals, most commonly the New Zealand White rabbit. This type of rabbit was chosen because of their large eyes, which are easier to inspect for any product-related damage.

Although testing to establish if a cosmetic or its ingredients will damage the eyes can now be performed using non-animal methods, they cannot differentiate the ability of a cosmetic to cause stinging of consumers' eyes. In reality, this effect can only be studied in real life, using human volunteers.

It is important to note that this type of testing is not performed on a regular basis. It is usually reserved for products for which there is a risk of them getting into the eyes. These products include those which are used on the face near to the eyes and baby washing products such as shampoos and body washes.

However, before the product is placed in the eyes of the volunteers, it must first be screened using a non-animal testing method to establish the highest concentration that will not cause any obvious damage when put into the eyes of the volunteers.

Once an appropriate highest concentration level has been selected, the product is then diluted to give a series of solutions of different strengths weaker than the starting concentration. The different solutions are then dripped into the volunteer's eyes in a stepwise manner starting with the weakest dilution.

In most cases, a similar benchmark product that is already being sold to consumers is also tested in parallel for comparative purposes. In all cases, the diluted products are dripped into only one of the volunteer's eyes, the second eye acting as an untreated control. The experiment is then stopped at the point where the volunteer starts to notice that their treated eye is stinging.

The results obtained are critically reviewed and sometimes analysed statistically, to arrive at the concentration that most commonly causes stinging of the eye to occur. This concentration can then be compared with the equivalent benchmark product to determine whether the new product performs better or worse than the benchmark.

The results of the study can also be used to support certain eye-related claims for the new product, such as 'kind to eyes' or 'no eye stinging'.

9.5.2 Hypoallergenic – Exactly What Does This Mean?

A statement that is often seen on the labels of certain products (particularly targeting consumers who have 'sensitive skin') is the 'hypoallergenic' claim. Although consumers may interpret this statement in different ways, the bias tends to be towards skin irritation. In strict medical terminology, this type of cosmetic will be less likely to cause skin allergies when compared with similar products that do not state the same claim.

However, unlike sun protection claims, for which there are official definitions and symbols, at the time of writing there are no equivalent definitions and symbols for this claim. This means that cosmetic brand owners who wish to make this claim on their products are free to define it in any way that they wish. This can result in a lack of clarity and hence further confusion for the consumer.

In response, the European Commission recognized that this situation is unhelpful to consumers. Consequently, it published a set of guiding principles that were aimed at giving advice to European cosmetic companies on how they should go about making products that claimed to be hypoallergenic.

Included in these principles was advice about the type of ingredients that consumers could reasonably expect to be excluded from products with a low potential to cause skin allergies. For example, the guidance suggests that substances that dermatologists and toxicologists report as being the cause of skin allergies should be avoided as ingredients in hypoallergenic products. Similarly, substances that are found to be potential skin allergens in experimental models should also be avoided, in addition to those which are classified in the same way for safe handling by factory workers and similar operatives.

In summary, these guiding principles are aimed at preventing interested and knowledgeable consumers from being exposed to ingredients that have the potential to cause allergic skin reactions.

However, this is exclusively an EU initiative, which means that cosmetics purchased outside the EU may not be manufactured with the same principles in mind. As a result, some non-EU products that claim to be hypoallergenic may in fact contain ingredients that are known to produce skin allergy reactions, which would disqualify them from being used in similar products in the EU. Therefore, consumers who purchase such products in the belief that they will not produce any skin allergies will be given a false sense of security.

9.6 POST-MARKET SURVEILLANCE – THE CUSTOMER'S STORY OF USING A PRODUCT

Have you ever wondered what cosmetic companies do if a customer informs them that they have experienced a skin reaction after having used one of their products? The answer is cosmetovigilance or post-market surveillance.

The cosmetic legislation in many parts of the world demands that cosmetic companies implement certain processes into their customer care procedures for dealing with customer complaints of a medical nature. These types of complaints could include customers who experience skin rashes, complaints of a medical nature or stinging of the eyes, for example. These processes and procedures are commonly known as cosmetovigilance, which mirrors an equivalent procedure in the pharmaceutical industry known as pharmacovigilance. It is partly for this reason that cosmetic companies in the EU, and increasingly in other areas of the world, have to include their address as part of the pack information for consumers. Very often these contact details will include a telephone number for customers to call, usually free of charge, to report problems they have experienced.

However, with rapid advances in electronic methods of communication, including social media, consumers now have many more options to contact a company to report a problem that has been experienced.

Having received a skin-related complaint from a customer, the company concerned should follow a set of standard procedures known as 'causality assessment', which are aimed at discovering whether or not the product was truly the cause of the customer's complaint.

The European Cosmetic Industry Trade Association, Cosmetics Europe, has published a scientific paper that describes a useful set of cosmetovigilance procedures and related terms for cosmetic companies. Although the use of these guidelines is optional for companies, their adoption will ensure that customer complaints of a medical nature are processed in a standard way. This will ensure consistency in the recording and reporting of customer complaints among the European cosmetic industry.

However, on some isolated occasions, the causality assessment will reveal that it was not the product itself that was at fault but rather the result of a customer's misuse of the product in question. An example of this is the mistaken use of a skin cleansing product, which is designed to be rinsed from the skin, as a skin moisturizing product, which is designed to be left on the skin. These two are fundamentally different product types since the cleanser will contain a blend of detergents, which will cause skin irritation and drying if left on the skin for prolonged periods of time. Detailed questioning of the customer, as part of the causality assessment process, should reveal that the customer misused the product. It could be that the misuse came about because of poor instructions for use or no instructions for use. The insight gained would be useful to the brand and should prompt them to review the usage instructions given to customers, which could help to prevent the reoccurrence of similar customer complaints in the future.

9.7 WHAT DOES ALL THIS MEAN TO ME?

In conclusion, contrary to the many myths and scare stories that are sometimes perpetuated on social media sites, consumers who use cosmetic products can have confidence that their products are not going to cause them any harm.

The EU has been at the forefront in improving standards of safety and quality for cosmetics sold in Europe. Indeed, as a compliment to this European philosophy, more and more countries around the world are adopting this model as their own.

It is not only safety standards that are being improved in the European model, European philosophy is also setting high standards for the effectiveness of the cosmetics we use every day. This means that in parallel with safety, customers can be assured that the products that they buy will actually do what is claimed!

FURTHER READING

EU Cosmetic Regulation, https://eur-lex.europa.eu/legal-content/EN/TXT/PDF/?uri=CELEX:02009R1223-20160812&from=EN.

US Cosmetic Ingredient Review, https://www.cir-safety.org.

EU Scientific Committee on Consumer Safety, https://ec.europa.eu/health/scientific_committees/consumer_safety_en.

Joint FAO/WHO Expert Committee on Food Additives, http://www.inchem.org/pages/jecfa.html.

European Commission, https://ec.europa.eu/growth/sectors/cosmetics/animal-testing_en.

EU Reference Laboratory for Alternatives to Animal Testing, https://ec.europa.eu/jrc/en/eurl/ecvam.

Cosmetics Europe Long Range Science Strategy, https://www.lrsscosmeticseurope.eu.

European Commission Technical Document on Cosmetic Claims, https://ec.europa.eu/docsroom/documents/24847.

W. M. S. Russell and R. L. Burch, *The Principles of Humane Experimental Technique*, 1959, Methuen, London.

G. Renner *et al.*, Cosmetics Europe Guidelines on the Management of Undesirable Effects and Reporting of Serious Undesirable Effects from Cosmetics in the European Union, *Cosmetics*, 2017, **4**, 1.

Safety Gate: the rapid alert system for dangerous non-food products, https://ec.europa.eu/consumers/consumers_safety/safety_products/rapex/alerts/repository/content/pages/rapex/index_en.htm.

CHAPTER 10

Myths and Scares – Science in Perspective

E. MEREDITH[*,a] AND R. POLOWYJ[b]

[a] Cosmetic, Toiletry and Perfumery Association, Sackville House, 40 Piccadilly, London W1J 0DR, UK; [b] IMCD UK Ltd, Times House, Throwley Way, Sutton SM1 4AF, UK
*Email: emeredith@ctpa.org.uk

Throughout this book, we have learnt about the strict legislative framework for cosmetic products, the work involved in ensuring that cosmetic products are safe, the in-depth structure of the skin, the science behind key ingredients and specific products and claim substantiation. Yet despite all of this and the wealth of information that cosmetic scientists have researched over many years, cosmetic and personal care products are often the subject of negative media coverage and there are many myths and scare stories circulating on the Internet. Above all, there seems to be confusion, and some concern, surrounding the use of 'chemicals' in general, including what they are, why they are used and whether they are safe – especially in cosmetic products. Media articles very often do nothing to dispel this so-called 'chemophobia', with scary headlines such as 'Is your skin

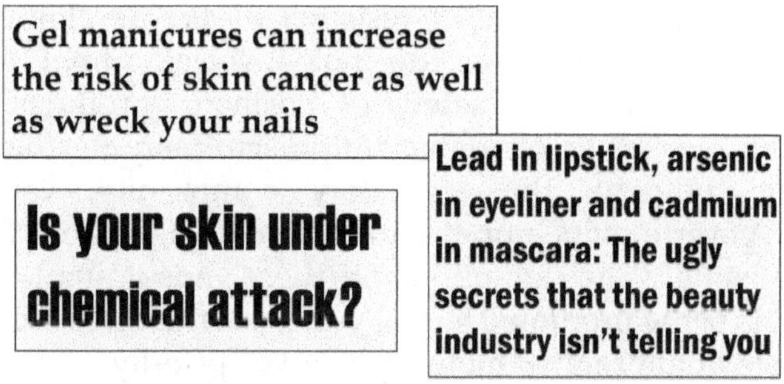

Myths and Scares – Science in Perspective 291

> Gel manicures can increase the risk of skin cancer as well as wreck your nails

> Is your skin under chemical attack?

> Lead in lipstick, arsenic in eyeliner and cadmium in mascara: The ugly secrets that the beauty industry isn't telling you

Figure 10.1 Media headlines.

under chemical attack?' undermining the safety of our cosmetic products (Figure 10.1).

This chapter provides factual and clearly laid-out information to help dispel the most common myths about cosmetic products and provides the science behind the industry's safe and innovative products.

10.1 ARE COSMETICS TESTED ON ANIMALS?

10.1.1 European Union (EU)

Cosmetic products sold in Europe are not tested on animals and this is true whether or not the product makes an 'animal friendly' claim. Animal testing of cosmetic products and cosmetic ingredients is banned in the UK and across Europe. Despite these bans having been in place for many years, the misconception still remains about cosmetic products and animal testing. The ban on animal testing of cosmetic products in the EU has been in place since 2004 and animal testing of cosmetic ingredients in the EU became illegal in 2009. In the UK, testing of cosmetic products has been banned since the late 1990s after a voluntary initiative by industry that led to all licences for testing cosmetic products to be withdrawn.

The bans apply to all cosmetic products and their ingredients sold in the UK and EU, whether they are made in Europe or anywhere else in the world. This means that no cosmetic products sold in the UK and EU have been tested on animals.

Not using animals does not mean that cosmetic manufacturers compromise on safety. The cosmetics industry has invested heavily in continuing research on the development of alternative approaches that do not involve animal testing, ensuring the continued safety of cosmetic products. Such research is the result of many multidisciplinary partnerships, including industry, the regulatory community, validating agencies, governments, non-governmental organizations (NGOs) and academia. The adoption of these non-animal testing methods is being promoted globally and international regulatory acceptance of alternative methods is a key priority.

You will have read about some of these methods in Chapter 9.

10.1.2 Global Challenges

Although several other areas around the world are seeking to introduce bans on animal testing for cosmetic products, it is the case that some countries still require animal testing of cosmetics under their own laws. However, companies, industry bodies, UK and EU regulators and other key stakeholders are working with the authorities from those countries, with scientific exchanges and training workshops, to explain why, based on the experience in Europe, animal testing of such products is not necessary to ensure safety.

10.2 HOW MUCH DOES THE SKIN ABSORB?

Skin is truly amazing. It is the largest organ of the body and it does a very important job – it keeps all of our insides inside! It also acts as a protective barrier against external factors. It has three main functions:

- Protecting our bodies from infection, *e.g.* due to bacteria or other microbes.
- Allowing us to feel what's around us so we can detect danger and pick things up with the right amount of force.
- Helping to regulate the body's temperature and keeping it constant.

Also, vitamin D is produced in the skin, stimulated by sunlight.

As you will have learnt from Chapter 5 on skincare, the outermost layer of the skin is called the epidermis. The cells you can see on the surface of the skin are constantly being shed. The skin renews itself by producing more cells at the base of the epidermis.

10.2.1 Myth – 60% of Everything You Put on Your Skin Is Absorbed

It is important to remember that the skin is a barrier, not a sieve. If the skin acted like a sieve, then when we have a bath we would absorb most of the water. If you were to run an experiment to test this theory and spread some petroleum jelly on your arm, just see how long it would take for 60% to be absorbed. You'll have plenty of time to read this book!

Whether something can be absorbed through the skin or not is complex. It is based on many different factors and it depends on the characteristics of each individual substance, it is not just down to size. The skin is effective against penetration, which is why even today most medicines have to be swallowed or injected and very few can be absorbed through the skin from topical patches.

Of course, some ingredients do penetrate the skin (you will have read more on this in Chapter 8) or may be ingested (from oralcare and lip products) or the skin may be broken (due to cuts, grazes, inflamed areas), which allows substances that would not normally be able to pass through the skin to gain access to the body. However, these are often readily broken down by the body and harmlessly processed and eliminated. They do not accumulate within the body to reach unsafe levels. Also, all of these elements will be addressed by the safety assessor and will be factored into the safety assessment that you read about in Chapter 9. The safety assessment takes into account any possibility of skin penetration, including as necessary any use on damaged skin, in addition to any possible ingestion or inhalation, to ensure that no harm is caused to you by using the cosmetic product.

10.2.2 Fact – Different Skin Types Have Different Barrier Functionality

It is often claimed that babies' skin is not as well formed as adult skin. It is true that baby skin is more delicate than that of adults and can be damaged by coarse fabrics or rough towels, for example. This is partly because baby skin is slightly thinner than adult skin (about 20–30%). It is also because skin responds to the environment and babies are making the transition from life in the womb to life in the outside world and therefore experiencing it for the first time. Baby skin also has a higher surface pH (a scientific measure of acidic or alkaline conditions).

However, it is not true to say that ingredients can be more easily absorbed through infant skin. Babies are born with skin that is very nearly complete in its barrier function and this further matures within the first 2–4 weeks after birth, providing an effective barrier to external and unwanted substances.

Personal care products made for babies and infants are formulated taking into account the delicate nature of baby skin; for example, they use milder cleansers and low levels of fragrance and carefully control the pH to ensure compatibility with the skin.

10.3 SHOULD I AVOID CERTAIN INGREDIENTS?

You should avoid certain ingredients if you know that you have become sensitized to them. This means that your body reacts to an ingredient in a way that it never did before (usually through redness or irritation). This is also the case with allergies. These cases are specific to individual people and the same ingredient will be able to be used by others without experiencing any irritation or sensitization.

10.3.1 Why Are Some Products Labelled As Being 'Free-from' Certain Ingredients?

There has been a big trend in the personal care industry for 'free-from' claims that appear on products' packaging and company websites/social media. These claims relate to specific ingredients that brands claim their products do not contain. This type of claim can often be misleading as it portrays the ingredient in a negative way – some consumers see claims such as 'paraben free' or 'silicone free' and may assume that they are left out of the formula because they are bad for you, which cannot be the case as they must be safe as used in cosmetic products. Often, these types of claims are created purely for marketing purposes with no reliable science behind them. Any claim made by a cosmetic product must be substantiated and must not mislead the consumer.

10.3.1.1 European Union.
According to Commission Regulation (EU) No. 655/2013, claims on cosmetic products should conform to the following common criteria:

1. legal compliance
2. truthfulness
3. evidential support
4. honesty
5. fairness
6. informed decision-making.

Owing to the rise in 'free-from' claims and the misconceptions that can be perceived from them, the European Commission and EU Member States have published guidelines for 'free-from' claims that came into effect in July 2019. A few examples of these guidelines are listed in Table 10.1 against each of the common criteria.

Table 10.1 Examples of claims that cannot be made under the common criteria.

1. Legal compliance	'Free-from' claims should not be claimed if the ingredient is already prohibited for use in cosmetics
2. Truthfulness	If something is claimed on the product such as 'free from formaldehyde', it should not contain a material that releases formaldehyde (some ingredients convert into other materials on the skin so it's important that if you claim something isn't in the formula, the materials that are being used do not then convert into this)
3. Evidential support	The absence of certain ingredients cannot be claimed unless adequate and verifiable evidence is available
4. Honesty	'Free-from' claims should not be allowed when they refer to an ingredient that is typically not used in a particular type of cosmetic
5. Fairness	'Free-from' claims should not be allowed when they imply a negative message. For example, 'free from parabens' should not be allowed as certain parabens are safe to use
6. Informed decision-making	'Free-from' claims with similar meaning may be used when they allow an informed choice to a specific target group or groups of end users. An example is the claim 'free from alcohol' in a mouthwash where the product is intended for family use or 'free from animal-derived ingredients' for products intended for vegans.

Some of the most common ingredients that are claimed as 'free-from' are parabens, sulfates and silicones, but what are these materials and what myths have caused people to avoid them?

10.3.2 Parabens

'Paraben' is a broad term used to describe the chemical structure of a specific group of ingredients. There are numerous different types of parabens, such as methylparaben and ethylparaben, which are often used as preservatives in cosmetics. A preservative is something that keeps our products free from bacteria and mould so they are safe for us to use over a period of time. Parabens can be manufactured or are also found abundantly in Nature (see Box 10.1).

10.3.2.1 Myth – Parabens Cause Breast Cancer. This myth originates from a small study conducted in 2004 that found parabens in breast cancer tissue. This does not mean that parabens cause cancer and Cancer Research UK explains:

Finding parabens in tumours is a far cry from saying that they cause breast cancer. To do that, scientists would need to compare levels in breast tumours to 'safe' thresholds, to levels in healthy body cells or to levels in healthy people without cancer.

BOX 10.1 QUESTION

Thinking about the following items, what do they all have in common?

- blueberries
- beer
- vanilla
- green tea
- medicine
- cosmetics

The answer is they all contain parabens.

Many studies exist on the safety of parabens and overwhelmingly the scientific data show that parabens used in cosmetics are not harmful.

10.3.3 Sulfates

'Sulfate' is a term given to a substance that contains a specific chemical group based on sulfur and oxygen, which has the symbol SO_4^{2-}. They are used for their cleaning properties and are very effective surfactants (surface-active agents). Sodium lauryl sulfate (SLS) and sodium laureth sulfate (SLES) are two of the best foaming and cleansing agents on the market and are commonly used in body washes, shampoos and hand washes.

10.3.3.1 Myth – SLS/SLES Cause Cancer. SLS and SLES have been found safe by the US Cosmetic Ingredient Review panel (CIR) and the American Cancer Society has discredited the 'SLS/SLES cause cancer' myth. When SLS was reviewed by the Australian Government's chemical assessment scheme (NICNAS), it was also found safe. In Europe, the safety of SLS and SLES has not been questioned by the EU or Member State regulatory authorities. Also, these ingredients have not been listed as causing cancer or suspected to cause cancer by any of the organizations around the world that determine such properties.

Prolonged contact with high concentrations of surfactants can be irritating for skin. SLS is often used in dermatological studies in investigations relating to irritation. In these studies, high concentrations of SLS are left on the skin and covered for several days before irritation occurs. This is not how SLS and SLES are used in cosmetic products. They are used at low concentrations in rinse-off cosmetics, often with other materials such as moisturizing ingredients, to ensure that such irritation does not occur.

10.3.4 Silicones

Two-thirds of the Earth's crust is made up of silicon, making it the most abundant material on Earth after oxygen. It is the fourteenth element in the Periodic Table and ironically sits just below carbon (of which we humans are made). This position in the Periodic Table means that silicon behaves and performs more similarly to carbon than any other material and, because of this, it is commonly used in medicine, implants and injections.

When used in cosmetic products, 'silicone' is a general name used to describe the chemical structure of ingredients that contain silicon, oxygen, carbon and hydrogen. There are hundreds, if not thousands, of different types of silicones, ranging from solids to semi-solids, but most commonly they are used in liquid form. In cosmetics, they are mainly used because they can provide an excellent skin feel in creams and lotions. For hair they can make it glossy and healthy looking.

10.3.4.1 Myth – Silicones Block the Skin's Pores and Cause Build-up on the Hair.

Silicones have been extensively used in skincare for decades, with studies showing that they are non-comedogenic (they don't block pores!). In haircare, the hair shaft, which is made of keratin, is negatively charged (anionic). If you put silicone on your hair and then wash your hair, the silicone just rinses off because silicones are usually non-ionic, meaning they don't have a positive or negative charge. Since there is no charge, there is no attraction between the silicone and the hair – they don't stick together.

To make a silicone or a natural oil stick to the hair, you need to add a positively charged (cationic) depositing agent to the formulation. The positive depositing agent is attracted to the negatively charged hair like a magnet, which then coats the hair and allows the oil to stay on the hair even after rinsing.

When 2-in-1 shampoos (a shampoo and a conditioner in one product) first came onto the market many years ago, the formula used very strong depositing agents. Because they were so strong, they stayed on the hair for a long time, even after several washes, which caused some product build-up to occur. This left the hair feeling heavy, dull and less shiny. Since that time, silicones have had a bad reputation for causing build-up, which is now not the case. Nowadays, 2-in-1 shampoos are much more advanced and use different depositing agents that are less attracted to the hair, meaning that build-up with continued use should not occur.

10.3.5 How Do I Know That the Ingredients in the Products I Use Are Safe?

It is unfortunately the case that some people can be allergic to commonly used substances. Each person is different and someone might find themselves allergic to ingredients that others use or consume without any problems. For example, many people can eat peanuts (groundnuts) and yet some cannot. This does not mean that peanuts are unsafe, it is the way in which one person's body reacts to them differently from another. The same can happen with cosmetic ingredients.

It is not possible to avoid all substances in cosmetics that might cause a rare allergic reaction in someone, just as we can't avoid all foods to which someone might be allergic. For cosmetic products, if you have been diagnosed as allergic to a specific ingredient it is important always to check the ingredients list on the packaging of the cosmetic you intend to buy or use to make sure that it does not contain any ingredients to which you are allergic.

Ingredients are not harmful when used by a scientist properly. Like anything, quantity is key and too much of anything (even too much water!) can cause negative effects on the body. This is the basis of toxicity (see Box 10.2).

When creating cosmetic products, scientists always assess the hazards and risks of ingredients before using them. The terms hazard and risk are sometimes confused with each other, but they mean very different things. A hazard is an unsafe characteristic that something has, whereas the risk is the chance of harm happening should you encounter (be exposed to) the hazard. There is an equation for determining risk:

$$\text{risk} = \text{hazard} \times \text{exposure}$$

Consider a lion as an example. If a lion is uncaged and roaming freely, it can be considered dangerous (a hazard) because it might attack you. If the lion is locked in a cage, the risk of it attacking

---

BOX 10.2 WHAT IS POISONOUS?

'All substances are poisonous: there are none that are not. The dose alone differentiates a poison from remedy.'
(Paracelsus, 1493–1541)

you is completely reduced. The lion is still a dangerous lion but you can safely be near it as long as you are outside the cage!

10.3.6 Do Cosmetics Contain Hormone-disrupting Ingredients?

You may hear claims about 'endocrine disruptors' being present in cosmetic products. This is not the case.

'Endocrine disruptor' is the term given to certain substances that can act as, or interfere with, human hormones in the body and lead to harmful effects.

Hormones act like a 'communication system' for the body. They are chemicals produced by the body as part of the endocrine system. There are many different types of hormones, such as thyroxine, insulin, oestrogen and testosterone, and each one has its own particular function within the body. Hormones affect our growth, how we react to situations, how we process sugar and how we develop sexually. In the EU, hormones (for example, progestogens, oestrogens and anti-androgens of steroidal structure) are prohibited from being present in cosmetic products.

Endocrine disruption is an emotive topic, but cosmetic ingredients are not endocrine disruptors. There is a wealth of scientific information that supports the safety of ingredients and nothing linking them to a decline in fertility or abnormal endocrine effects.

The World Health Organization (WHO) defines an endocrine disruptor as follows:

> *An endocrine disruptor is an exogenous substance or mixture that alters function(s) of the endocrine system and consequently causes adverse health effects in an intact organism or its progeny or (sub)populations.*

The crucial point is that to be considered as an endocrine disruptor the substance must produce adverse health effects in a whole body.

It is important to remember that just because something has the potential to mimic a hormone *in vitro* (when tested outside a living organism) it does not mean that it will disrupt the endocrine system *in vivo* (in the human body).

While it may be the case that certain substances may mimic some of the properties of our hormones or may, under experimental conditions, show a potential to interact with parts of the endocrine system, these conditions are not related to real life. For example, the UV (ultraviolet) filter benzophenone-3, used in some sun protection products, is 1.5 million times less potent in its oestrogenic effect than ethinyloestradiol, which is used in oral contraceptives. To put this into perspective, if aspirin were 1.5 million times lower in potency, we would need to consume more than 13 times our body weight of pure aspirin to treat a headache.

Many so-called 'endocrine disruptors' (actually endocrine mimics) are abundant in nature. We ingest them in the food we eat in concentrations many times greater than in cosmetics and personal care products. Endocrine mimics include phytoestrogens – oestrogen-like compounds found in plants. We eat these in foods such as cabbage, soya beans and sprouts, and no adverse health effects have been associated with these dietary exposures.

10.4 WHAT IS THE DIFFERENCE BETWEEN NATURAL AND SYNTHETIC INGREDIENTS?

There is no legal definition of 'natural' or 'synthetic' in the context of cosmetic products at the time of writing this book, but it is generally recognized that the term 'natural' means a substance that has been obtained directly from nature, be that something from a plant, a mineral or an animal. 'Synthetic' (sometimes called 'man-made') can be thought of as something that has been made in a laboratory.

It may be the case that certain manufacturing processes are applied to a material sourced from nature, which then means it could be classed as being synthetic. For example, petroleum jelly is considered a synthetic material, but if you were to trace its origin it derives from crude oil, which occurs naturally. The oil undergoes several manufacturing processes that result in a synthetic product.

It is also possible to re-create naturally occurring substances in a laboratory. The term 'nature identical' is often used to describe this process. These substances look and behave in exactly the same way as their naturally occurring equivalents.

As the natural cosmetic market grows, you may be asking 'Are natural ingredients safer for us than synthetic ones?'. For cosmetic products, the answer is no. The important thing to remember is that for cosmetic and personal care products, all ingredients used must be safe whether they are sourced from nature or from a laboratory. The body cannot determine whether an ingredient is from a natural source or is manufactured, and where it comes from has no bearing on how safe it is either. What is important is how much of the ingredient is used and in what way it is used. See Box 10.3.

The important thing to remember for cosmetic products is that no matter where an ingredient is sourced, whether from nature or a laboratory, it has to be safe.

Many ingredients can have long names and can seem scary. For example, 2-oxo-L-*threo*-hexono-1,4-lactone-2,3-enediol (Figure 10.3) might look like it shouldn't be used by humans. However, if we see its common name, vitamin C, we wouldn't have any concerns!

Similarly, the number of ingredients present in a product has no impact on its safety. Cosmetic and personal care products are very complicated to make and they often have a long list of ingredients to get them just right. Each ingredient within that list will have a specific role to play, from making the product work effectively, to making it smell and feel nice, to making it last for a satisfactory amount of time. As explained in Chapter 9, each one of these, individually and in combination, will have been rigorously assessed for safety before placing the product on the market. However, the safety assessor does not simply count

---

BOX 10.3 EXAMPLE

Essential oils, which are considered a natural product, are extracted from a plant and they may contain high levels of substances (allergens) to which consumers may be allergic or sensitized. When synthetic fragrances are recreated in the laboratory, scientists can isolate a specific scent molecule and produce it as a pure ingredient (Figure 10.2) – so without or with fewer allergens. This means that consumers with allergies may be able to use them too.

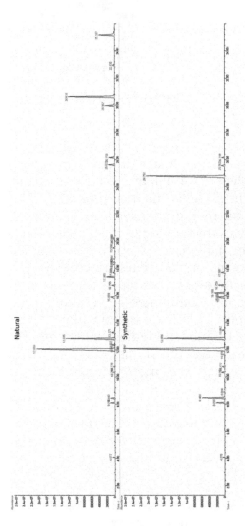

Figure 10.2 Recreating a fragrance in the laboratory from analysing the natural extract [image courtesy of Dr Garry R. Dix (Regulatory Expert, CPL Aromas) and Tim Gage (Senior Perfumer, CPL Aromas)].

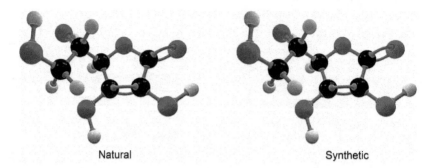

Figure 10.3 Vitamin C molecules, natural *versus* synthetic (© Alfred Pasieka/ Science Photo Library).

the number of ingredients and use that as a guide to safety. Instead, they use their knowledge and skill that come from many years of training to assess the safety of the final product.

What if we had to list all of the chemicals in an apple? Figure 10.4 lists the substances that are found in an apple. If we had to label apples with an ingredient list we might want to question their safety, but no-one thinks apples are unsafe!

The naming of chemicals follows set protocols that relate to their structure. The names can sometimes be very long and complicated! To help simplify this for cosmetic ingredients and to avoid people having to know ingredient names in different languages, many years ago the industry agreed on a common naming system called the International Nomenclature for Cosmetic Ingredients, or INCI. The same ingredient names are used in every European country and most countries worldwide. Although these names can sometimes appear complex, this is necessary to identify each ingredient precisely and the name is usually simpler than the full chemical or botanical name.

10.5 DO COSMETICS POLLUTE THE OCEANS?

Ingredients are subject to legislation for other chemicals that looks at their environmental aspects and safety. Companies also look at environmental effects such as how much energy is used, how much water is involved, how much waste is produced and

Figure 10.4 The composition of an apple (Image courtesy of Sense About Science).

what damage is caused when obtaining raw materials and turning them into cosmetic ingredients and products – the sustainability and impact of the product. Not only do cosmetic products and their ingredients need to be safe for human health but the safety of the ingredients and products for the environment and sustainability factors must also be considered.

10.5.1 Do Cosmetics Contain Plastic Microbeads?

Plastic microbeads are small, solid, plastic particles used to exfoliate or cleanse in some rinse-off personal care products. The small plastic beads were originally selected for use as exfoliating or teeth-cleaning agents because they are clean, safe, can be produced to be a uniform size and have no sharp edges to scratch the skin.

Plastic microbeads are one type of microplastic. Cosmetic products are an extremely small source of microplastic contributing to the problem of plastic marine litter. In 2016, the environmental consultancy Eunomia estimated this to be 0.29% and highlighted the main sources of microplastic marine litter.[1] However, as an environmentally responsible industry, the cosmetics industry acted voluntarily to remove plastic microbeads from products ahead of bans introduced in several countries around the world.

10.5.2 Are Cosmetics a Cause of Microplastics?

There is sometimes confusion about the terminology used in this important topic and the terms 'plastic microbead' and 'microplastic' are often confused with each other, but they mean different things (see Box 10.4). Plastic microbeads are the hard, solid, plastic beads that may have been used in a variety of

---

BOX 10.4 SOME DEFINITIONS

Plastic microbeads – Hard, solid, plastic beads used in a variety of products.

Microplastic – Tiny, solid, plastic particle or fibre found as litter in oceans and other waterways.

products, including cosmetics and personal care products, such as scrubs. Plastic microbeads from cosmetic products are a very small fraction of microplastic marine litter. Microplastic refers to any type of tiny, solid, plastic particle or fibre found as litter in oceans and other waterways.

Microplastic most often starts as larger pieces of plastic debris, such as plastic packaging, cigarette filters, car tyres or synthetic fabric, that breaks down into tiny pieces over time. These particles and fibres measure 5 mm or less in diameter and do not dissolve in water. Microplastic that started as larger litter is called 'secondary microplastic', whereas particles that are intentionally developed as small plastic particles are called 'primary microplastic'.

Marine plastic litter is a global issue and microplastic originates from a variety of sources.

10.5.3 Why Is Plastic Packaging So Often Used to Package Cosmetics?

Plastic is an important and useful material. However, the disposal of plastic is a major global problem that affects all industries and consumers. The cosmetics industry is working to be a part of the wider solution, through research and activities in this important area. For a tangible benefit to the environment, action is required by both industries and consumers to promote responsible disposal and work towards a circular economy (Box 10.5, Figure 10.5).

Packaging has important functions. Its main aim is to protect its contents from spoiling, so protecting the consumer and enabling the consumer to store and use the product safely over time. It must be strong enough to:

- withstand the rigours of an automated manufacturing process (filling and final packaging for transport);

BOX 10.5 CIRCULAR ECONOMY

Resources (*e.g.* packaging materials) are kept in use for as long as possible through recovery, regeneration and re-use rather than throwing something away.

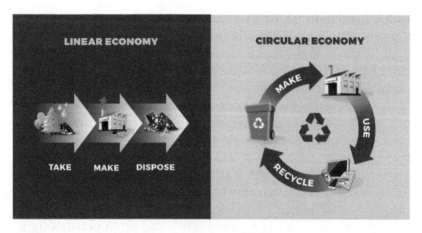

Figure 10.5 Linear *versus* circular economy (© Petovarga/Shutterstock).

- withstand transport and storage;
- fit on shelves in supermarkets or other retailers;
- look attractive to the people who might buy it; and
- stay looking good and be robust throughout its life, which for cosmetics may be several months or even years.

Cosmetic packaging must be able to be labelled with specific legally required information, including a list of ingredients and, where necessary, how to use the product safely.

There are laws in place to ensure that packaging must be able to be recovered, which includes recycling. Cosmetics manufacturers try to strike the right balance between a product that is protected for use over several months or even years and minimizing packaging. Packaging manufacturers have been innovating in the area of sustainable packaging for some time and are working on solutions to increase the sustainability of packaging. Packaging suppliers look at new methods and technologies for making packaging, such as more environmentally friendly printing inks, improved design of bottles, new machinery and using more recycled materials. Using recycled plastic is to be encouraged, although it is important that there is a sufficient amount of good-quality feedstock to maintain the source of recyclable plastic while ensuring that what is used is safe for cosmetic products and consumers and does not contain unwanted or unsafe impurities.

10.5.4 Are Sunscreens Damaging to Coral?

Information from Cancer Research UK has highlighted that UV radiation is the third largest contributor to cancer cases in the UK.[2] It is therefore not surprising that global public health authorities recommend the use of sun protection products as part of a sun-safe regime to help reduce the risk of skin cancer.

There is no proven link between the use of UV filters in sunscreens and damage to coral in our seas. However, unfounded claims about the safety and efficacy of sun protection products could undermine confidence in these products in general, which could have serious health implications if consumers choose to avoid the use of any sunscreen if trust in them has been diminished.

The study upon which the stories and the legislation banning certain UV filters in some parts of the world are based does not reflect what happens in nature. The study was carried out in a laboratory under artificial conditions, and did not replicate what is happening in the real world. In particular, though, the study does not make a link between the use of UV filters in suncream and damage to coral in our seas.

The deterioration of coral reefs around the world is, of course, a major concern. The factors that contribute to it are very complex and it is widely acknowledged within the scientific community that extreme climate events and warming sea temperatures are the major contributing factors. Interestingly, regions of the world where the most severe coral bleaching has occurred do not match where human population density is higher, which suggests there may not be a link with human activity, such as sunscreen use.

10.6 WHY ARE COSMETICS SO IMPORTANT?

Many cosmetics and personal care products are essentials in everyday life as part of daily routines that keep our bodies clean, smelling fresh and looking good. There are many functional benefits, such as handwashing with soap can reduce the risk of illness, the use of toothpaste reduces the occurrence of dental caries and regular use of sunscreens against UV rays, as one part of a sun-safe regime, helps in the battle against skin cancer.

There is also a fun side to cosmetics that helps us show our personal style and reflect our personality.

There are numerous studies linking the use of cosmetic products with self-confidence and self-esteem. In particular, the report *Consumer Insights 2017*, published by Cosmetics Europe, the European Personal Care Association, provides a snapshot at the European level of consumer perceptions of the cosmetics and personal care industry (Figure 10.6). The research was carried out by an independent research company, polling over 4000 consumers across 10 EU Member States.

The report shows how consumers consider that cosmetics and personal care products play an important role in building self-esteem and enhancing social interactions every day. On the

Figure 10.6 *Consumer Insights 2017* (Image courtesy of Cosmetics Europe).

importance of products enhancing self-esteem, 80% of consumers identified cosmetics and personal care products as important or very important in building up self-esteem. This perception was reflected across all age ranges, from young millennials to older age groups, all relying on products that matter to them to help enhance how they feel in their daily lives.

How important to you are the products you use? Do you have a special product that you just couldn't do without?

10.7 HAVING CONFIDENCE IN COSMETIC PRODUCTS

Cosmetic and personal care products are enjoyed by millions of people all over the world, contributing to healthy lifestyles, wellbeing and confidence in appearance.

It takes a team of scientists to develop, manufacture and market each cosmetic product. From concept to final product, the sequence will include basic biological research into specialist ingredients, formulation development and checking, efficacy testing, scaling up to manufacturing from laboratory development, packaging, further efficacy testing, safety assurance and regulatory compliance. Each and every step involves many different scientific disciplines. In fact, safety is the number one priority of the cosmetics industry, with manufacturers recognizing their responsibilities and often going beyond their legal obligations when it comes to product safety.

Responsible manufacturers invest their reputation in their brands, building trust with loyal customers and relying on customer satisfaction for success. Fundamentally, that trust is based on delivering products that are safe, effective and of high quality. After all, if you are not happy with a product, you won't buy it again!

Before reading this book, did you ever stop to wonder about how the cosmetics and personal care products we use each day are made, who is involved and how it all comes about?

It is hoped that the information presented throughout the book, and in particular in this chapter, has helped to explain just how much the cosmetics industry is steeped in science and how safety is considered at every stage of developing a cosmetic product. By dispelling some common myths, we hope that you will have confidence in your cosmetics and personal care products and take for granted that they will be of good quality, effective and safe.

REFERENCES

1. C. Sherrington, Eunomia, *Plastics in the Marine Environment*, Available from: https://www.eunomia.co.uk/reports-tools/plastics-in-the-marine-environment/ (Accessed 7 February 2020).
2. K. F. Brown, *et al.*, *Br. J. Cancer*, 2018, **118**, 1130.

APPENDIX 1

Career Opportunities in the Field of Cosmetic Science

Now you have read this book, you may be interested in considering a career in the field.

Generally, a cosmetic scientist will have completed a science degree at university, either directly relating to cosmetics or in another discipline such as chemistry, biology or pharmaceutical chemistry. One very well known cosmetic scientist started with a degree in nuclear physics!

There are other pathways – cosmetic scientists can start by joining the industry with pre-degree qualifications and, gaining experience along the way, work through the career ladder, possibly completing a degree or a diploma in cosmetic science (the Society of Cosmetic Scientists runs a Diploma in Cosmetic Science course, accredited by the Royal Society of Chemistry). There are many different career paths within the cosmetic industry and, almost without exception, these will require working in teams with a range of scientific expertise. Some of the main functions where qualifications in science generally or cosmetic science specifically can be an advantage include the following:

- **Research and Development** – This is often a laboratory-based role where new materials are researched and tested before being developed into a cosmetic product. Equally it could be

---

Discovering Cosmetic Science
Edited by Stephen Barton, Allan Eastham, Amanda Isom,
Denise McLaverty and Yi Ling Soong
© The Royal Society of Chemistry 2021
Published by the Royal Society of Chemistry, www.rsc.org

from research into hair, skin or oral microbiology. Whether it is ground-breaking research or clever application of existing technology, often R&D scientists will apply for patents to protect their discoveries. The R&D role is vital in areas such as optimizing ingredient performance, reducing energy utilization in production or improving the product's benefits or safety.

- **New Product Development** – This can be both a laboratory- and an office-based role. This role is at the heart of all new developments and involves creating an idea for a new cosmetic product and ensuring that the science is fit for the intended purpose. These team members know how long the R&D chemists need to work in the laboratory to create the product. They also know what type of, and how much, testing is required to ensure that a safe and effective product gets to market.
- **Production** – Just about every product you use will be made in bulk – in batches varying from as little as 5 kg up to 2000 kg. Some processes are automated, some are manual. All require an understanding of the chemistry, the flow properties, the melting points, the flammability and many other aspects of the ingredients and final products.
- **Packaging** – We do not deal with the chemistry of packaging materials in this book – there are other sources dedicated to this aspect. However, we should acknowledge that the pack is often an integral part of the product. This is demonstrated very well in the case of a mascara, where the applicator design and product flow need to be carefully considered together during development. Some cosmetic scientists use their knowledge of the product chemistry in the packaging department – the two disciplines complement each other.
- **Quality Assurance** – All products and ingredients used in cosmetic need to be reliable. You expect your shampoo to foam the same way every time you use it. Cosmetic scientists know that even small variations in the quality of an ingredient can be the difference between a successful product and something that ends up washed down the drain. They use 'specifications' for ingredients and products. For many materials, ones that have been used for many years, this can be as simple as a physical description – a white powder

soluble in cold water producing a salty taste describes sodium chloride. Increasingly, specifications are based on measurements made with sophisticated analytical chemistry methods. These methods help assure aspects such as purity, absence of toxins and the presence of active molecules. Another factor that is growing in importance is proof of origin. The science here is becoming equally sophisticated with new methods such as DNA profiling or tagging being introduced in a few cases.

- **Testing** – Many organizations will test their own products in ways described in Chapter 8. Others will send products for testing in specialist testing organizations. In both cases, scientists skilled in the measurement of the physical, chemical and biological properties are important participants in ensuring that the products do what they claim to do and remain stable over their shelf life.
- **Toxicology** – Every cosmetic product must be safe for consumers to use. A trained professional toxicologist will determine how chemical substances affect living organisms. Armed with this knowledge, published data and results from other testing methods, sound decisions can be made on whether a product is suitable for the market.
- **Perfumery** – To be a perfumer can be challenging as there are so few perfumery schools around the globe. A cosmetic science degree covers a few perfumery units, which opens up an entry point into the perfumery industry. As we show you in Chapter 7, this knowledge is also important in making body lotions, shampoos and even depilatories smell pleasant during use.
- **Regulatory** – Cosmetic regulations are continuously changing, so a regulatory expert should know the applicable laws in depth to ensure that their products do not break any regulations, which are different for each country. Many of these regulations are based on science and the chemistry of the product – so training in cosmetic science is essential.
- **Technical Sales** – There are hundreds of cosmetic ingredient manufacturers all around the world creating innovative scientific ingredients. Many use distributors in different countries working directly with local brands and contract manufacturers to supply all the materials they need in one

place. Technical sales people in the distributors need to know exactly how each material works on the skin, hair or oral cavity and within a formulation. Again, a cosmetic science qualification is crucial to provide this service to the product developers.
- **Publishing/Journalism** – Scientific and specialist commercial publishers are interested in new research within the cosmetic science field. Letting people know new developments in cosmetics does need an understanding of cosmetic chemistry. With the rise of social media, there has been a surge in cosmetic bloggers with huge followings who inform consumers about products they like to use.
- **Teaching** – The popularity of cosmetic science continues to rise, as universities add this course to their prospectus. This creates more opportunities to go into teaching and sharing existing knowledge with new students.
- **Marketing** – In the past, marketing individuals would study a marketing degree to enter this type of role, but increasingly cosmetic scientists are joining this field and frequently starting to create their own product ranges. Technical knowledge can help to create a brand image and a unique selling point, be that through which ingredients they use, the effects the customer is looking for, the type of cosmetic products they sell or which packaging the consumer will prefer.
- **Entrepreneurship** – Having the knowledge to create cosmetic products is a skill for life and can lead you into creating your own brand and your own products. Many individuals go on to create successful companies and consultancies to help others produce the cosmetics they like.

Anyone wanting to find out more about education or careers in chemistry or cosmetic science should visit https://www.scs.org.uk and https://www.rsc.org.

Subject Index

References to tables and charts are in **bold** type

α-hydroxy acids (AHAs), 149–50, 235–7
accords, 202, 205
acetone, 221
acetyl hexapeptide-8, 234
acid mantle, 129
acid violet, **6**
acne, 58, 111, 115, 122–3, **124**, 230, 237, 246
Actinomyces spp., 79
aftershave, 149
agglomerates, 36, 174–7
aggregates, 66, 174, **175**
agitation, 25
air fresheners, 197, 199
air/water interfaces, **21**
alcohol free, **295**
aldehyde notes, 205, 209, **218–9**
alkyl ether sulfates, 29, 39–40, **41**
alkyl polyglucosides (APGs), 39–40, **41**
alkyl sulfates, 27–8, 35, 38–40, **41**
alkyl sulfoacetate, **41**
allantoin, 122
allergens, 131, 286, 302
allergic reactions, 5, 162, 189, 278, 286, 299, 302
allergies, 8, 189–90, 286, 294, 302
almond notes, **213**
almond oil, 132, 147
aloe vera, 36, 132
alopecia
    androgenic, 61
    areata, 61
alumina
    see aluminium oxide
aluminium
    borosilicate, 172
    chlorohydrate, 260
    oxide, 93–4, 103, 165
    silicate, **6**
alveolar bone, 81
alveoli, 76
amber, 203, 205
American Association of Textile Chemists and Colorists, 166
American Cancer Society, 297

American Dental Association (ADA), 91
amidoalkyl betaine, **41**
amine fluoride, 92
ammonia, 28, 79, **152**, 188–9
ammonium lauryl sulfate (ALS), 38–9
amphiphilic agents, 21, 26
amygdala, 196
amylase, 78
anagen phase, 56, 62
androgenic alopecia, 61
androgens, 61
    anti-androgens, 300
anethole, 103
animal testing, 32, 277, 291–2
animalic notes, 203, 205, 208
aniseed, 103
anomalous trichromats, 159
anthocyanins, **245**
anthranilates, 209
anti-acne, 230
anti-ageing, 9, 181, 227
anti-androgens, 300
anti-caking, 93
anti-dandruff, 66, 253
anti-foaming agents, **135**
anti-inflammatory, 100
anti-sensitivity, 89, 92, 94–7, 101–2, 104–5
antibodies, 78–9
anticorrosion, 244
antimicrobial peptides (AMP), 115
antioxidants, 6, 14, 120, 225, 241, 243–5, 258
apatite
    fluorapatite, 92
    hydroxyapatite, 77, 83, 90, 95, 98, 102, 104–6
    hydroxyfluorapatite, 83, 90
    strontium apatite, 95
apocrine gland, 55
apple, 14, 157–8, **206**, 229, 242, 304, **305**
areata alopecia, 61
arginine, 83, 95
arrecetor pili muscle (APM), 55
Arrhenius equation, 210
artificial sunlight, 270, 283, **284**
ascorbic acid, 229
ascorbyl glucoside, 229
ascorbyl palmitate, 229
asparagus, 88, 245, **245**
Aspergillus brasiliensis, 248, **249**
astringents, 150
attachment loss, **81**
aubergine, **245**
autoimmune conditions, 61
axon, **193**
Ayuvedic, 101, 253
azo pigments, 163, **164**

β-hydroxy acids (BHAs), 235–6
baby skin, 271, 293–4
bad breath, 88
bakers yeast
    see Saccharomyces cerevisiae
baking soda
    see sodium bicarbonate
bamboo, 99
barrier function, 115, 130–1, 293–4
basal cell, **193**
base notes, 200, 202–4
basement membrane (BM), **55**
Baur, Albert, **214**
beard, 2, 46, 123
beer making, 7, 248, 296
beeswax, 146, 178

beetroot, 103, 104, 162
bell peppers, **245**
bentonite clay, 179
benzoates, **6**
benzoic acid, **6**, 8, 250
benzophenones, 240
    benzophenone-3, 301
    benzophenone-4, **6**
bergamot, 205, 211, 219
best before date, 272
beta lipid layer, 49
bicuspids, 75
bio-active glasses
    see bioglass
bioactive, 232
bioavailability, 227
biofilm
    see plaque
bioglass, 95, 102
bioglitter, 181
biotin (B_7), 230
birch, 220
bisabolol, 122
bismuth oxychloride, 169, 182
black iron oxide (CI 77499), 166–7
black tea, 106
blackcurrants, 206
blackheads, 123, 150
bleach, 208
bleaching, 50, 88, 97–100, 188, 309
bleeding effect, 165
blood
    clots, 52
    dental, 76, 87
    infection, 246
    red blood cell count (RBC), 32
    stream, 125, 257
    supply, 55, 61
    vessels, 113, 120, 127, 179
    white blood cells, 79
bloodstream, 125, 257
bloom, 183, 200
blow-drying, 69–70
Blue 1 Al Lake (CI 42090), 165
blue covarine, 98
Blue No. 1, **103**
blueberries, 296
blusher, 173
body butters, 129
Body Shop, The, 254
boiling points, 219–21
bones crushed, 93
botanicals, 7, 36–7, 304
botulinum toxin (Botox), 234
boundary lubrication, 67
box braids, **44**
Brazilian keratin treatment, 72
bread making, 248
breast cancer, 296
bricks, 93, 123, 130
British Dental Association, 91
broccoli, **245**
bromelain, 98
brown iron oxides, 167
brushite
    see dicalcium phosphate dihydrate
Brussels sprouts, **245**
bulge region of hair, **55**, **62**
bulking, 93
burning skin, 119–20, 282–4
butternut squash, 245
butylmethoxydibenzoylmethane, 240
butylparaben, **250**
Butyrospermum parkii, **6**, 7–8

cabbage, **245**, 301
cade oil, 220
    juniper notes, **203**, 220
caffeine, 61, 153, 254
cake foundations, 180
calcium aluminium borosilicate, 172
calcium carbonate, 93–5, 98, **103**
calcium chloride, 152
calcium fluoride, 94
calcium oxalate, 95
calcium peroxide, 99
calcium pyrophosphate, 90, 93, 98
calcium sodium phosphosilicate (CSPS), 95, 102
calculus, 80–1, 87, 89, 98, 100, 102, 104, 107
camouflage, 46
cancer, 72, 244, 309
    American Cancer Society, 297
    breast cancer, 296
    Cancer Research UK, 296, 309
    mutagenic changes, 99, 237, 278, 296–7
    skin cancer, 119, 237–8, 282, 309
    tumours, 296
Cancer Research UK, 296, 309
Candida albicans, 248
candles, 199
canine teeth, 75
capric
    see caprylic
caprylic
    chains, 28
    glucosides, **41**
    glycerides, 183
    triglycerides, 132
caprylyl glycol, **6**, 7
carbomers, 6, 103, 145, 178
carbon dioxide, 221–2, 253
carcinogens
    see cancer
carmine dye, 162
carminic acid, 162
carotenoids, **103**, 104, 162, **214**, **228**, **245**
carrageenan gum, 103, 145
carrots, 228, **245**
castor oil, 36, 147, 176
catagen phase, 56, 62
causality assessment, 287–8
celery, **245**
cell cultures, 256, **278–9**
cell membrane complex (CMC), 25, 30, 49–51, 60
cell monolayers, 278, **279–81**
cellulite, 153
cellulose, 11, 103
    hydroxyethylcellulose, 103
    microcellulose, 150
cemento-enamel junction, **81**
cementum, 76–7
Centella asiatica, 122
cera alba, 178
ceramides, 130
cetearyl alcohol, 139, 146
cetearyl glucoside, 137, **139**
cetyl alcohol, **6**, 7–8
chalk
    see calcium carbonate
chamomile, 36
Chanel, Coco, 43
charcoal, 99
chassis, 5, 9
cheek piercing, 85

chelation, **6**, 97
chemophobia, 290
chewing gum, 92
chlorhexidine, 100–1, 104, 106
chlorine dioxide, 99
chlorophyll, **103**, 104, 162
cholesterol, 130
chroma, 160–1
chromium oxide (CI 77288), 167
chromogens, 79, 87, 98–100
chromophores, 127
chypre fragrance family, 205
cigarettes, 79, 87–8, 119
    filter debris, 307
cilia, 193
cinnamon, 103
Ciste labdanum, **203**, 205
citric acid, 6, 8, 235, 260
citronella, **214, 218**
citronellal, **214, 218**
citrus notes, 203, 205, 209, 216, **217–8**, 219
clays, 151, 180
    bentonite, 103, 179
    hectorite clay, 103
    montmorillonite, 179
    organoclay, 179
clenching, 85
Cleopatra, 253
clove oils, 103
Cnut, King, 70
coacervates, 66
coalescence, 180
cocamidopropyl betaine (CAPB), **6**, 33–5, 38–40, **41**, 103, 137
cochineal beetle, 162
coco glucoside, **41**
cocoa butter, 131–2, 183

coconut, 255
    husks, 99, **255**
    Lactobacillus spp., **255**
    meal, **255**
    medium chain triglycerides (MCT), **255**
    oil, 28, 35, 39–40, 101, 132, **255**
    trees, 40
    water, **255**
coenzyme, 230
coffee, 87
coffee grinder, **177**
cold-pressed expression *see* expression extraction
collagen, 77, 111, **112**, 117, 122, 126, 229–30, 233–4, 238, 281
colour blindness, 159
colour creation
    lipsticks, 162, 165
    sunlight, 163
colour fastness, 270
combustion, 243
comedo
    closed, 123
    open, 123
comedones, 123
    microcomedones, 123
compromised immune system, 267
concealers, 180
concrete, 221
cones, 156
connective tissue sheath (CTS), 55
consumer perceptions, 24, 39, 310
consumer safety, 4, 276, 288
contraceptives, 301
coral reefs, 309

corn, 146, **245**
corneocytes, 111, 123, 125, 128–30
corneodesmosomes, 111, **125**
cornflour, 143
corrosion, 241, 244
cortex, 46–7, 49–50, 53, 59, 69, 195, 230
cortical cells, 47, 49–50, 53
Corynebacterium spp., 115
Cosmetic, Toiletry and Perfumery Association (CTPA), 17, 37
Cosmetics Europe, **4,** 277, 287, 310
cosmetovigilance, 263, 273, 287
COSMOS, 39
cotton, 152
coumarin, 205, **213**
courgettes, **245**
cradle cap, 58
cradle to cradle, 255
creaming, 180
creative teams, 198–9, 224
cribiform plate, **193**
critical micelle concentration (CMC), 25, 30, 49–51, 69
crude oil, 301
crushed china, 93
curling irons, 59
currants, **245**
customer complaints, 287–8
customer confidence, 174
Cutibacterium acnes, 58, 115, 123–4
cuticle, 47, 49–52, 59–60, 185, 187–8
    endocuticle, 52
    epicuticle, 51–2
    exocuticle, 52

cuttlebone, 93
cyclic terpenes, **217**
cyclopentasiloxane, 147
cyst, 124

dandruff, 58, 66, 115, 253
de Predis, Ambrogio, **44**
decamethylcyclopentasiloxane, 148
decorative cosmetics, 3, 173
decyl glucoside, **41**
deforestation, 40
delivery systems
    lipsticks, 258
    stratum corneum (SC), 258–9
delta lipid layer, 49
dementia, 107
demineralization, 77, 82–3, 90, 102, 105
dendrites, 111, **193**
dental abrasion, 84–5
dental attrition, 84–5
dental blood, 76, 87
dental caries, 82–4, 88–92, 99, 101, 104, 309
dental erosion, 84–5
dentine tubules, 85, **86,** 91, 94–6, 102, 105
dentures, 84
deodorants, 206, 230, 257, 259–60, 273
deoxyribonucleic acid (DNA), 48, 119, 238, 247, 249–50, 315
dermal papilla (DP), **55**
dermatitis, 58, 115
dermatologists, 8, 109, 119, 183, 286, 297
dermis, 111, **112,** 114, 117, **118,** 120, 127, 131, **282**

Subject Index

desquamation, 111, **113**, 117, **118**, 123
dewberry, 254
diabetes, 107, 244
dicalcium phosphate dihydrate, 98, **103**
diesel fuel, 169
diethanolamine (DEA), 35–6
diffuse reflection, 127, **127–8**, **128**, 157–8, 158, **173**
dihydroacetic acid, **6**
dihydrotestosterone, 61
dimethicones, **6**, 66, 138, **139**, 147, **148**
dipeptide diaminobutyroyl benzylamide diacetate, 234
dipeptides, 233–4
dishwashing liquids, 197, 199, 203
disodium cocoyl glutamate, **41**
disodium lauroyl sulfosuccinate, **41**
dispersant agents, 93
distearyldimonium chloride, **139**
distillation extraction, 216, 219–22
DMDM hydantoin, 250
Dodge, F.D., **214**
dry distillation
    see distillation extraction
dye precursors, 187–9
dyers woad
    see woad
dyes
    melanin, 185–6, 188

eau de toilette (EdT), 206–7, 209
eczema, 115, 122, 267
eggs, 15, 32, 88, 93, 168, 228

Egyptian times, 93, 184, 253
elastin, 111, **112**, 117, 229, 233–4, 238
electromagnetic spectrum, 117, 155, 157–60, 169, **170**, 237–8
electrostatic, 30–1, 67–8
emotions and smell, 195–6
encapsulated, 181, 258
endangered plant species, 222–3
endocrine disruptors, 300–1
endocrine system, 300–1
endocuticle, 52
enfleurage, 221
environmental stressors, 148
Eomaia scansoria, **45**
eosin dyes, 162
epicuticle, 51–2
epidermis, 111, **112–4**, 117, **118**, 120, 125–7, 129–30, 257, 281–2, 292–3
erythema, 119, 149, 237–9, 282–3
Escherichia coli, **247**
esterification, 171
esters, 27, 132, 146–7, 176, 178, 208, **218**
    methylionones, **218**
    polyglyceryl esters, 37, 137
    retinyl esters, 228
ethanal, 209
ethanol, 7, 142, 206–7, 221
    diethanolamine (DEA), 35–6
    monoethanolamine (MEA), 28–9, 189
    triethanolamine (TEA), 28, 137
ethinyloestradiol, 301

ethoxylates, 29, 135, 137, **139**
ethoxylation, 29, 135, 137, **139**
ethyl acetate, 221
ethyl alcohol (ethanol), 7, 104, 142, 206–7, 221, 272–3
ethylene glycol, 171
ethylene glycol distearate (EGDS), 171
ethylene glycol monostearate (EGMS), 171
ethylene oxide (EO), 29, 32, 265
ethylhexyl
    glycerine, 250
    methoxycinnamate (octinoxate), 240
    palmitate, 132
ethylparaben, 296
EU Cosmetics Directive, 88, 91
EU Reference Laboratory (EURL), 277
EU Scientific Committee on Consumer Safety, 276
eucalyptol, 103
Eugenia caryophyllus see clove
eukaryotic cells, 248
eumelanin, 53, 186
Eunomia environmental consultancy, 306
European Centre for the Validation of Alternative Methods (ECVAM), 277
European Cosmetic Industry Association, 277
European Cosmetic Industry Trade Association, 287
European Personal Care Association, 310
evaporation, 60, 130, 202–3, 207, 220

evening primrose oil, 131
evolutionary biology, 45
exfoliation, 150, 235
exocuticle, 52
expectoration, 101
expression extraction, 216, 219
expression lines, 234
extracellular matrix, 111
extrinsic skin ageing, 117
eyeshadows, 173–4, 181–2, 210

fabric softeners, 204
facial oils, 147, 151
fatty alcohols, 27–9, 67, 132, 137, 139, 146
Fédération Dentaire Internationale (FDI), 90–1
ferric ammonium ferrocyanide (CI 77510), 167
ferric salts, 95
fertilizers, 11
fibrin, 52
fibroblasts, 111, **112**
fish, 243
fish liver oil, 228
fish scales, 168
fissures, 79
Fittig, Wilhelm Rudolf, **213**
Fitzpatrick scale, 120, **121**
fizzy drinks, 84
flaking, 129, 183
flavonoids, **245**
flocculates, 174, **175**
flossing, 81
fluorapatite, 92
fluorescence, 157
fluoristan, 91
fluorosis, 90, 92, **93**
fluorphlogopite, 172
flyaway hair, 67–8

foaming agents, 23–4, 39, **41**, 98, 104, 150, 163, 297
folic acid, 231
follicles, 44–6, 53–8, 60–2, 73, 111, 123, 125, 153, 172
Food and Agricultural Organization, 276
forensics, 77
formaldehyde, 72, **295**
fougères fragrance family, 205, **213**
fragrance families, 204–5
    chypre, 205
    fougères, 205, **213**
    oriental, 205
fragrance houses, 196–8, 208, 211, 215
fragrance oil, 35, 206
free from claims, 295
free from sulfates, 296
free radicals, 58, 60, 119–20, 228, 241, 243–4
friction tests, 67
fructose, 78
fruit juices, 84–5, 105
Fusobacterium nucleatum, 79

Gantrez, 100
garlic, 88
gas chromatography, 64, 222–3
gelling agents, 103
genotoxicity, 278
geranium, 205
geranyl acetate, **218**
ginger, 222
gingival crevice, **81**
gingivitis, 80–1, 92, 100–1, 107
glass/liquid interfaces, 24
glass/water interfaces, 21
glazing, 183

glitters, 168, 181
    bioglitter, 181
glomerulus, **193**
glutamates, 39–40
    disodium cocoyl glutamate, **41**
    sodium stearoyl glutamate, **139**
glycation, 117
glycerin, 102–4, 131, 250
glyceryl
    laurates, 37
    oleates, 37, **139**
    stearates, **6**, **139**
glycine, **233**
glycol, 5, 102
    caprylyl glycol, **6**, **7**
    ethylene glycol, 171
    ethylene glycol distearate (EGDS), 171
    ethylene glycol monostearate (EGMS), 171
    glycolic acid, 153, 235–6
    hexylene glycol, **6**, **7**
    pentylene glycol, 250
    poly ethylene glycol (PEG), 6, 36, 102, 137–8, **139**
    propylene glycol, 102, 131
    thioglycolate, 72
glycolic acid, 153, 235–6
glycoproteins, 78, 111, 117, 131
glyoxylic acid, 72
Gobley, Nicolas-Théodore, **214**
golden syrup, 30
good manufacturing practice (GMP), 268
gourmand notes, 203, 205
Gram negative, 79, 246–7
Gram positive, 79, 246–8

grapefruit juice, 105
green chemistry, 137
green tea, 296
grinding, 85
groundnuts, 299
growth factors, 78
growth phase
    see anagen phase
guanine, 168
guar gum, 145, 178
guar hydroxypropyltrimonium chloride, 66
gumline, 75, 77, 79, 85
gums
    carrageenan, 103, 145
    guar, 145, 178
    locust bean, 145
    vegetable, 144–5
    xanthan, 103, 145, 180

Haarmann, Wilhelm, **214**
hair browning, 59
hair bulb, 55–6, 62
hair density, 53, 60–1
hair diameter, 53, **54**
hair frizz, 48, 69
hair greying, 61–2
hair relaxation, 72
hair stiffness, 48, 69
hair thinning, 54, 57, 60–2
hammer mill, 176–7
hard wax, 178
Hawthorne effect, 96
hay, 213
headache, 301
headroom, 251
headspace analysis, 222–3
heart disease, 107, 244
heart notes, 200, 202–3
heat styling, 59, 69

hectorite clay, 103
Hegman gauge, 184
heliotrope flower, **213**
heliotropine, **213**
helmet head, 70
hemoglobin, **127**
henna, 184, **185**, 190
hens egg test - chorioallantoic membrane test (HET-CAM), 32
herrings, 168
hexylene glycol, **6**, 7
hippocampus, 196
holistic, 255
homogenizer, **175**
homosalate, 240
honey, 253
hue, 160–1
human volunteers, 234, 238, 283–5
humectants, **6**, 89, 102–4, 131, 133, 178, 250
humidity stress, 210
hyaluronic acid, 111, 117, 131, 151, 258
hydrodynamics, 67
hydrogen peroxide, 99, 104, 187
hydrogen sulfide, 88, 168
hydrophilic, 20, 35, 60, 63, 125, 133, 135–6, **140**
hydrophilic-lipophilic balance (HLB), 35, 135–6, 139
hydrophobic, 20–1, 60, 63, 98, 139, **140**, 148, 220
    see also lipophilic
hydroxy acids
    α-hydroxy acids (AHAs), 149–50, 235–7
    β-hydroxy acids (BHAs), 235–6

hydroxyapatite, 77, 83, 90, 95, 98, 102, 104–6
hydroxyethylcellulose, 103
hydroxyfluorapatite, 83, 90
hydroxypropyltrimonium chloride, 66
hygroscopic, 130–1, 230
hyperkeratinization, 123
hyperpigmentation, 236
hypoallergenic, 285–6
hypoallergenic products, 286
hypodermis, **112**, 113, **114**
hypothalamic, **195**

imidazolidinylurea, 250
imidazoline types, 33, **41**
immune system, 55, 61, 115, 124, 238, 267
immunoglobulins, 78
implants, 298
in vitro testing, 32, 238, 277–8, **279**, 284, 300
incisor teeth, 75
indigo, 159, 185, **186–7**
indoles, 209
infections, 83, 110, 115, 123, 129, 246, 248, 264–8, 292
inflammation, 80–1, 85, 100–1, 106, 123–4, 230, 243–4
infrared light (IR), **157**
ingestion, 90, 183, 293
inhalation, 72, 293
inner root sheath (IRS), 55
inorganic pigments, 162, 166–8, 176, 180
    black iron oxide (CI 77499), 166–7
    brown iron oxides, 167
    chromium oxide (CI 77288), 167
    ferric ammonium ferrocyanide (CI 77510), 167
    manganese violet (CI 77742), 168
    red iron oxide (CI 77491), 167
    ultramarine blue (CI 77007), 168
    yellow iron oxide (CI 77492), 167
insoluble azo pigments, 163
insulin, 300
interfaces, 24–5
    air/water, **21**
    glass/liquid, 24
    glass/water, 21
    liquid/air, 21, 24–5
    liquid/solid, 21
    liquid/surface, 30
    oil/dirt, 25
    oil/water, 133, **140**
    surface, 21
intermediate colours, 160
intermediate filaments (IFs), 47–8, 69, 72
International biodiversity laws, 255
International Federation of Cosmetic Chemists, 17
International Nomenclature of Cosmetic Ingredients (INCI), 5–8, 10, 12, **41**, 137–8, 139, 166, 215, 304
International Union of Pure and Applied Chemistry (IUPAC), 27
intrinsic skin ageing, 117
ionones, **214**
    methylionone, **218**

iron blue
   see ferric ammonium ferrocyanide
iron oxides, 167, 171, 180
   black iron oxide (CI 77499), 166–7
   brown iron oxides, 167
   red iron oxide (CI 77491), 167
   yellow iron oxide (CI 77492), 167
irritancy, 32, 39, 64
irritancy score, 32
Isatis tinctoria. see woad
isoprene, 216, **217**
isopropyl palmitate, 132

Jacobson's organ, **195**
Japan, 3
jasmine, 221
jellies, 173
Joint Expert Committee on Food Additives (JECFA), 276
jojoba oil, 132, 254–5
juglone, 185
juniper notes, **203**
   cade oil, **203**, 220

keratin, 47–8, 61, 72, 77, 123, 129, 298
   hyperkeratinization, 123
   keratinocytes, 111, **112**
keratin associated proteins (KAPs), 47–8, 69, 72
keratinocytes, 111, **112**
keratosis pilaris, 153
ketones, 208–9, **214, 218**

L*a*b* system of colour, 161
laboratory animals, 276–8, 284

lactic acid, 79, 83, 153, 235
Lactobacilli spp., 79, 250–1, **255**
Lactobacillus acidophilus, 250
lakes, 164–6
   Blue 1 Al Lake (CI 42090), 165
   Yellow 5 Al Lake (CI 19140), 165
lamellar gel, 139, 146
lamellar phase, 67
lanolin, 131
Laura Marshall Award Society of Cosmetic Scientists (SCS), 17
lauric acid, 101
lauryl alcohol, 27–8
lauryl betaine, 33
lauryl glucoside, **41**
lavender, 205, 208–9, **213**
lawsone, 185, **187**
lead arsenate, 168
lead carbonate, 169
leather notes, 203, 205
lemon, 14, **84, 214, 217,** 219
lemongrass, **218, 218–9**
lemons, 14, 84, **217,** 219
leukaemia, 72
light boxes, 159
limbic system, 195–6
limestone
   see calcium carbonate
limonene, 209, **217**
linear terpenes, **217**
linoleic acid, 131
lipid barrier, 32, 37, 130
lipophilic, 20, 35, 37, 133, 135–6, 146, 254
lipophobic, 20
   see also hydrophilic
lipoproteins, 258

liposomes, 258
lipsticks
    colour creation, 162, 165
    colour defined, 155
    cosmetic textures, 177, 183
    delivery systems, 258
    pigments, 174, 176
    product safety, 275
    skin texture, 147
liquid/air interfaces, 21, 24–5
liquid/solid interfaces, 21
liquid/surface interfaces, 30
liver, 110, 132, 228–9, 278
locust bean gum, 145
Louis XIV, King, 71
lubrication, 58, 66–7
lubricity, 5, 148
lustre, 168–9, 171
lutein, **245**
lycopenes, **245**
lye, 208
lymphocytes, 123

magnesium aluminum silicate, **6**
magnesium ascorbyl phosphate, 229
magnesium peroxide, 99
magnesium silicate, 182
magnesium stearate, 183
magnets, 12–3, 13, 30, 130, 167, 207, 224, 298
    electromagnetic spectrum, 117, 155, 157–60, 169, 237
Maillard reaction, 87
Malassezia spp., 58
manganese violet (CI 77742), 168

marine litter, 306–7
marine notes, 203
mascara, 267
massage oils, 147
mattifying, 179
mechanical insults, 46, 50
mechanical stiffness, 50
medium chain triglycerides (MCT), **255**
melanin
    cosmetic textures, 179
    dyes, 185–6, 188
    eumelanin, 53, 186
    hair damage, 60
    hair diversity, 53
    hair greying, 61–2
    peptides, 233
    pheomelanin, 53, 186
    product claims, 281–2
    skin care products, 127
    skin function, 111
    skin texture, 117, 120
    UV filters, 239
    vitamins, 229
melanocytes, 61–2, 111, **112**, **118**, 120, 281
menopause, 122
menstrual cycle, 123
Mentha piperita
    see menthol
Mentha spicata
    see spearmint
menthol, 103, 152, **217**
mercaptans, 88
mercuric chloride, 168
methyl salicylate, 103
methylionones, **218**
methylisothiazolinone, 250
methylparaben, 296
Mexican lorry driver's skin, 119

micelles, 24–5, 30–1, 34–6, 63–4, 67
microbeads, 150, 306–7
microbiome, 58, 114
microcapsule, 257–8
microcellulose, 150
microcomedones, 123
microcrystalline wax, 178
microencapsulation, 258
microplastics, 181, 306–7
microwaves, 237
Mielk, W.H., **213**
milk, 29–30, 35, 129, 145, 228, 235, 253
mineral oil, 13, 132
mineral water, 84–5
minimum erythemal dose (MED), 283
minocycline, 87
mint, **217**
mites, 115
moisturization, 129–31, 148–53, 179, 262
monochromats, 160
monoethanolamine (MEA), 28–9, 189
monoterpenes, 215, **217**
monoterpenoids, 215
montmorillonite clay, 179
mortar model, 125, 130
moulds, 50, 246, 248–50, 264, 296
    Aspergillus brasiliensis, 248, **249**
mouth jewellery, 85
mucilage, 145
musk notes, 203–5, 212, **214**, 254
mutagenic changes, 99, 237, 278, 296–7

myrcene, **217**
myristyl chains, 28
myristyl myristate, 146

Nagoya Protocol, 255
nail varnish, **173**, 177, 184
National Industrial Chemicals Notification and Assessment Scheme (NIC-NAS), 297
natural moisturizing factors (NMFs), 130–1
nature-identical versions, 254, 261, 301
neroli essential oils, 220
nerve endings, 86, 96
New Zealand White rabbit, 284
newborns, 37
Newtonian behavior, 142–3, 147
niacin (B_3), 226, 230
niacinamide, 230–1
nicotine, 125, 258
nitro musk, 212, **214**
Nobel Prize, 48, 194, 224
nodules, 124
non-animal testing, 277, 285, 292
non-comedogenic, 298
non-governmental organizations (NGOs), 276, 292
non-Newtonian behavior, 142–3
north light, 158
nutshells, 99
nylon, 182

oak nut galls, 185
oakmoss, 205
obesity, 107
octinoxate, 240
octocrylene, 240

Subject Index

octyldodecanol, 132
odontoblasts, 76–7, 86
odorant molecules, 194
odorant receptors, 193–4, 194
odour fatigue, 194
odour stability, 211
oesophagus, 248
oestrogens, 122–3, 300–1
    ethinyloestradiol, 301
    phytoestrogens, 301
oil pulling, 101
oil-loving
    see also lipophilic
oil/dirt interfaces, 25
oil/water interfaces, 133, **134–5**, 139, **140**, 173, 179–80
Olea europaea, **6**, 7
olfactive themes, 199–200, 202, 205–7, 215, 221
olfactory sensory neurons, 193–4
olfactory system, 192–6, 214
olive oil, 7, 99, 132, 167, 208, 227–8
olive pits, 99
opacifying, 93
optical nerve, 156
oranges, 85, **217**, 219, 221
orangutans, 40
organoclays, 179
oriental fragrance family, 205
orthodontic appliances, 84, 92
Oryza sativa, 178
osmotic pressure, 85–6
outer root sheath (ORS), 55, 62
over-the-counter (OTC), 3, 86, 91, 94, 96–7
ox hooves, 93
oxidants, 241, 243
oxidative stress, 62, 148, 241, 243

oysters, 168
ozone, 60, 237

p-phenylenediamine (PPD), **187**, 189
pain studies, 96
palm kernel oil, 28, 39
palm oil, 40, 136
palmitamidopropyltrimonium chloride, **139**
palmitic acid, 131, 228
palmitoyl tetrapeptide-7, 234
palmitoyl tripeptide-1, 233–4
panthenols, **6**, 7, 230
panthothenic acid (B_5), 226, 230
papain, 98
papaya, 98, **245**
paprika, 162
papules, 123–4
paraben free, 294, **295**, 296
parabens, 8, 245, 250, 294, **295**, 296–7
    butylparaben, **250**
    ethylparaben, 296
    methylparaben, 296
parotid duct, **78**
parotid gland, **78**
Parquet, Paul, **213**
patch tests, 32
patchouli, **203**, 205
pattern baldness, 61
Pauling, Linus, 48
peanuts, 299
pearlescent effects, 168–9, 172
pearlescent pigments, 158, 168–9, 171, 174–6, 180–1, 183
peas, **245**
peat, 99

pellicle, 78–9, 86–8, 91, 93–4, 98, 103
pellicle cleaning ratio, 94
pentaerythrityl tetraisostearate (PTIS), 178
pentylene glycol, 250
peppercorn, 222
peppermint, 103
period after opening (PAO), 272
periodontal disease, 80, **81**
periodontal ligament, 76
Perkin, William Henry, **213**
perlite, 98
permanent waving, 71, 73
peroxide treatments, 60, 88, 99–100, 106, 189, 228
peroxides
    calcium peroxide, 99
    hydrogen peroxide, 99, 104, 187
    magnesium peroxide, 99
pesticides, 11
petroleum jelly, 132, 293, 301
phagocytes, 123
pharmacovigilance, 287
phenoxyethanol, **6**, 7–8, 250
phenylbenzimidazolesulfonic acid, 239
pheomelanin, 53, 186
phospholipids, 258
photoreceptors, 156
phytoestrogens, 301
piercing, 85
pilosebaceous unit, 54
pine bark, **214**
pineapples, 98
pits enamel, 79, 83
placebo, 96
plaque, 79–84, 87–9, 92–3, 99–102, 106–7

plexus of Raschkow, 76
polarity, 12–3, 126, 207
pollution, 120, 148, 230
poly ethylene glycol (PEG), **6**, 36, 102, 137–8, **139**
polydimethylsiloxanes, 66, 147
polyglyceryl esters, 37, 137
polyglyceryl-4 caprate, 37
polyhedron foam, 33
polyphosphates, 98
polyquaternium, 66
polysorbate, 36, 137, **139**
post-market surveillance, 263, 273, 287
potassium pyrophosphate, 98, 104
potatoes, 11, 146, **245**
power toothbrushes, 80
prebiotic ingredients, 116
precarious lesions, 83
premolars, 75
preservative efficacy test (PET), 268–9, 272, 311
primary particles, 174, **175**, 176
primary surfactants, 29, 33–9, **41**, 42
principle colours, 160
pro-vitamin, 230–1
probiotics, 116, 251
progestogens, 300
prokaryotic cells, 247
Prophyromonas gingivalis, 79
Propionibacterium acnes
    see *Cutibacterium acnes*
propylene glycol, 102, 131
protofilaments, 47
protozoa, 79
Pseudomonas aeruginosa, 115
Public Health England, 91
pulp, 76–7, 83, 85–6

Subject Index

pumpkin, **245**
purple onions, **245**
purple potatoes, **245**
pustules, 123–4
pyridoxine (B_6), 230
pyrogallol, 185
pyrogenic, 93
pyrrolidonecarboxylic acid (PCA), 131

quince seeds, 145

rabbit, 284
radiance enhancing, 179
radio waves, 237
radiographs, 83, 237
rainbow, 53, 157
rainforest, 40
rancidity, 14, **242**, 244, 270
rapeseed, 40
raspberry, 206
recycled materials, 220, 308
Red 30 (CI 73360), 166
Red 36 (CI 12085), 166
Red 6 Barium Lake (CI 15850:2), 165
Red 7 Calcium Lake (CI 15850:1), 165
red blood cell count (RBC), 32
red iron oxide (CI 77491), 167
Red No. 40, **103**, 104
regression phase
    see catagen phase
Relative Dentine Abrasivity (RDA), 94, 97, 99
remineralization, 77, 82, 90, 96, 102, 104–5
resting phase
    see telogen phase
retina, 156

retinal, 228
retinoic acid, 228–9
retinoids, 126
retinol, 149, 228–9, 257
retinyl esters, 228
retinyl palmitate, 228
retinyl stearate, 228
rheology modifiers, 89, 103, 137, 141–7, 225
rheometer, 143
ribbon blender, 176
riboflavin (B_2), 230
ribonucleic acid (RNA), 249
rice, 146, 178
rice bran, 178
risk assessment, 263, 273, 275
risk free, 275
risk-benefit considerations, 275
Roddick, Anita, 254
rods, 156
rollup hypothesis, 64
root lift, 61, 69
rose, **203**, **214**, **218**, 220–1
rose absolute, 221
rosemary essential oils, 220
Roundtable on Sustainable Palm Oil (RSPO), 40
rust, 241, 244

Saccharomyces cerevisiae, 248, **249**
safety assessment, 190, 273–4, 276, 293
safety assessor, 274–7, 293, 302
safety data sheets (SDS), 276
salicylic acid, 235–6
saliva, 77–8, 82–5, 88, 95, 105, 267
salivary components, 78–9, 83

salivary glands, 78, 80
Salix nigra bark extract, **6**, 7
salt curve, 31
sandpaper, 98
sardines, 168
scalp massage, 61
scanning electron microscope, 95, 128, **170**
scrubbing machine, 106
seaweed, 103
sebaceous gland (SG), 55–8, 65, 111, **114**, 115, 123, 126, 129
    pilosebaceous unit, 54
seborrheic dermatitis, 58
secondary surfactants, 31–9, **41**, 42
sedimentation, 179–80
sensorial properties, 9, 142, 174, 177
serotonin, 237
sesame oils, 101
sharks, 92, 132, 273–5
shea butter, 7, 132
shear thickening, 142–3
shear thinning, 142–4, 146
shelf-life, 88, 251, 265, 269, 272, 315
shellfish, 168
Shiff's base, 209
silica-strontium deposits, 95
silicone free, 294, 296
silicone oil, 60
siloxanes, 147
    cyclopentasiloxane, 147
    decamethylcyclopentasiloxane, 148
    polydimethylsiloxanes, 66, 147
skin allergies, 286

skin barrier, 32, 58, 111, 130, 148
skin cancer, 119, 237–8, 282, 309
skin care products
    melanin, 127
    stratum corneum (SC), 125–7
    sunlight, 149
skin functions, 111, **112**, **113**, 114
skin microbiota, 114–6, 259
skin reactions, 286–7
skin texture
    lipsticks, 147
    melanin, 117, 120
skin tolerance, 263
slightly soluble azo pigments, 163
smoke
    see cigarettes
snail shells, 93
Society of Cosmetic Scientists (SCS)
    Laura Marshall Award, 17
sodium ascorbyl phosphate, 229
sodium bicarbonate, 93, 98, 101
sodium carbonate, 189
sodium citrate, **6**, 98
sodium coco sulfate, 39, **41**
sodium cocoamphoacetate, 33, 40, **41**
sodium fluoride, 90, 92
sodium hexametaphosphate, 94, 99
sodium hydroxide, **6**, 8, 28, 72, 137
sodium laural sulfate (SLS), 27–8, 32, 38–9, 103, 297

sodium laureth sulfate (SLES), **6**, 27, 29, 30–5, 38–9, **41**, 63, 297
sodium lauroyl methyl isethionate, 38, **41**
sodium lauroyl sarcosinate, **41**
sodium lauryl sulfoacetate, **41**
sodium metaphosphate, 93
sodium methyl cocoyl taurate, 103
sodium monofluorophosphate (SMFP), 91–2, 94
sodium percarbonate, 99
sodium polyacrylate, 145
sodium pyrophosphates, 98
sodium saccharin, **103**, 104
sodium stearate, 137, **139**
sodium stearoyl glutamate, **139**
sodium trimetaphosphate, 98–9
soft-focus effect, 172
solar simulator, 283
solubilizing agents, 13, 36, 42, **135**
soluble azo pigments, 163
solvent distillation, 221
solvent extraction, 216, 221–2
sorbitan oleate, **139**
sorbitan stearate, **139**
sorbitol, 6, 102–4, 131
soya beans, **265**, 301
spearmint, 103
spectrum, 100, 160
    electromagnetic, 117, 155, 157–60, 169, **170**, 237–8
    see also UV radiation
specular reflection, **127**, 157–8, 158, 173, 179
spice notes, 203
spinach, **245**
split ends, 50, 59
spoilage, 116, 265, 267–8, 273
sports drinks, 84
spreadability, 178
squalane, 131–2
squalene, 132
stability testing, 211, 263, 269–73
stannous chloride, 101
stannous fluoride, 91–2, 94
Staphylococcus aureus, 115, 246, **247**, 248
Staphylococcus epidermidis, 115
star rating system, 239, 283
starch, 78–9, 145–6, 180, 182, 267
steam distillation, 220–2
steareth, **139**
stearic acids, **6**, 137, 171
stem cell biology, 56, 256, 261
Stevia rebaudiana, **103**, 104
stomach acid, 84
straightening irons, 59, 69
stratum corneum (SC)
    delivery systems, 258–9
    hair structure, 50
    hydroxy acids, 236
    moisturization, 129–31
    skin care products, 125–7
    skin function, 111, **112**, 113, 114
    skin types, 117, **118**, 123
Streptococci spp., 79
stress, 61–2, 79, 142, 144
    humidity, 210
    oxidative, 62, 148, 241, 243
stretch marks, 153
stripping effect, 32–3
strontium apatite, 95
strontium salts, 95

styling polymer, 70
subcutis, 281
subgingival plaque, 79
sublingual
    ducts, 78
    gland, 78
submandibular
    duct, 78
    gland, 78
sucralose, **103**, 104
sucrose, 78, 139
sucrose laurate, **139**
sugar cane, 235
sulfur trioxide, 27
sun protective factor (SPF), 119, 148–9, 238–40, 262, 283–4
sunburn
    see erythema
suncream, 262, 278, 309
sunflower oils, 101, 227–8
sunlight
    colour creation, 163
    hair damage, 60
    hair structure, 50
    odour stability, 211
    product claims, 280, 282–4
    skin absorption, 292
    skin care product types, 149
    stability testing, 270
    vitamins, 229
sunscreens, 3, 119, 238–40, 309
supercritical fluids, 221, 253
surface interfaces, 21
surface tension, 17, 22–3, 64, 103
surface-active agents, 21, 26, 103, 297
sustainability, 38–40, 306, 308

sweat, 111, 259–60
sweat glands, 45, 55, **114**, 129, 260
sweet almond oil, 132, 147
sweet potatoes, **245**
synthetic fabric debris, 307
synthetic fragrances, 302
synthetic silicone, 137, 147

tadpole shape, 19, 63, 67, 133
talc, 180, 182, 265
tangerine notes, **203**
tanning skin, 119–20
tapioca, 146
tar, 220
tartar
    see calculus
tea, 87, 106, 254
    black tea, 106
    green tea, 296
telogen phase, 56
temple vipor, 234, **235**
temporal cortex, 195
terminal hairs, 46, 56
terpenes, 208, 215–6, **217**
    cyclic terpenes, **217**
    linear terpenes, **217**
    monoterpenes, 215, **217**
terpenoids, 216, **217–8**
    monoterpenoids, 215
testosterone, 122–3, 300
    dihydrotestosterone, 61
tetracycline, 87
tetrapotassium pyrophosphates, 102
tetrasodium pyrophosphates, 102
thalamus, **195**
The Cosmetics Regulation (EC), 166

The Safety Gate Rapid Alert System, 269
thermal insulation, 46–7
thiamin (B$_1$), 230
thioglycolate, 72
thiols, 88
third molars
    see wisdom teeth
thixotropic system, 179
thyme, **217**
thyroxine, 300
Tiemann, Ferdinand, **214**
tiger grass, 122
time-controlled release, 258
titanated mica pearls, 169
titanium dioxide, **103**, 161, 169, **170**, 180, 182, 239
tobacco
    see cigarettes
tocopherols, 227
tocopheryl acetate, **6,** 228
tocotrienols, 227
tomatoes, **245**
toners, 150, 164–6
    Red 6 Barium Lake (CI 15850:2), 165
    Red 7 Calcium Lake (CI 15850:1), 165
tongue piercing, 85
tonka beans, 213
tooth abscess, 83
tooth extraction, 83
tooth stains, 87–8, 98
top notes, 200, 202–3, 205, 222
topical patches, 293
tottles, 271
toxicity, 10, 168, 274, 276–8, 299
toxicology, 286, 315
trans-epidermal water loss (TEWL), **118,** 130–1, 151

transdermal drug patches, 125, 258
treatment masks, 151–2
Treponema pallidum, 248, **248**
tribology, 67
triclosan, 100
tridecyl ether, 36
triethanolamine (TEA), 28, 137
triethyl citrate, 260
trinitrotoluene (TNT), **214**
tripeptides, 233–4
triple-roller mill, 176
true pigments, 164–6
    Red 30 (CI 73360), 166
    Red 36 (CI 12085), 166
tuberose, 221
tumours, 296
turmeric, 87, 162
tyre debris, 307

UK Society of Dyers and Colourists, 166
ultramarine blue (CI 77007), 168
ulus oil, **6**
unique selling point (USP), 181
universal solvent, 22
urea, 83, 130–1, 152
    imidazolidinylurea, 250
US Cosmetic Ingredient Review, 276, 297
UV exposure, 59–60, 119, 190
UV filters, 163, 181, 237, 239, 261, 270, 309
    benzophenone, **6**, 240, 301
    ethylhexyl methoxycinnamate (octinoxate), 240
UV radiation, 60, 117–20, 148, **152**, 211, 238–40, 309

UVA light, 117, 119, 237–40, 280–3
UVA protection factor (UVA-PF), 239–40
UVB light, 117, 119, 237–9, 280–3
UVC light, 237, 280

vanilla, 205, 209, **213–4**, 270, 296
varnish, **173**, 177, 184
Vaseline
    see petroleum jelly
vegetable gums, 144–5
vegetable oil, 132, 136, 146–7
vellus hairs, 46
vessels
    blood, 113, 120, 127, 179
Victoria, Queen, 212
violets, 214, **218**
viruses, 79, 114–5, 249
VITA shade guide, **97**, 106
vitamin A, 228–30, **245**, 257
    retinoic acid, 228–9
    retinol, 149, 228–9, 257
    retinyl esters, 228
    retinyl palmitate, 228
    retinyl stearate, 228
vitamin B complex, 78, 230
    biotin (B_7), 230
    niacin (B_3), 230
    niacinamide, 230–1
    panthenols, **6**, 7, 230
    panthothenic acid (B_5), 230
    pyridoxine (B_6), 230
    riboflavin (B_2), 230
    thiamin (B_1), 230

vitamin C, 78, 151, 229–30, 302, **304**
    ascorbic acid, 229
    ascorbyl glucoside, 229
    ascorbyl palmitate, 229
    magnesium ascorbyl phosphate, 229
vitamin D, 237, 292
vitamin E complex, 227–8
    tocopherols, 227
    tocopheryl acetate, **6**, 228
    tocotrienols, 227
vitamin K, 231
volatile compounds, 79, 193, 202, 207, 216, 219, 222
vomiting, 85
vortex, 175

waglerin-1, 234
walnuts, 185
washing powder, 197–8
waste, 243, 255–6, 304
water-in-silicone (w/si), **139**, 173, 179
water-loving, 20–1, 24, 27, 30, 33
    see also hydrophilic
water/oil interface, 133, **139**, 173
watermelon, **245**
waterproof properties, 148, 258
weathering, 59, 158, 183, 238, 254
wetting agents, 22, 25–6, **135**, 180
whale oil, 254
white blood cells, 79
white musk, 254
whiteheads, 123

Subject Index

whitening teeth, 87–9, 97–100, 104, 106–7, 240
WiFi, 237
wigs, 71
willow bark, 7
wine making, 7, 87, 248
wintergreen, 103
wisdom teeth, 75, 104
witch hazel, 150
woad, 185, **186**
wood ash lye, 208
woody notes, 203–5, **217**, 219
wool, 131, **186**, 187–8
workhorse, 5, 232
World Dental Federation *see* Fédération Dentaire Internationale (FDI)
World Health Organization (WHO), 91, 276, 300
wrinkles, 117, **118**, 120, 128, **152**, 158, 172, 229, 234, 238

X-rays, 83, 237
xanthan gum, 103, 145, 180
xerostomia, 78
xylitol, 92, 102, **103**, 104

yeasts, 58, 79, 246, 248, 250
    Candida albicans, 248
    Saccharomyces cerevisiae, 248, **249**
Yellow 5 Al Lake (CI 19140), 165
yellow iron oxide (CI 77492), 167
ylang ylang essential oils, 220
yoghurt, 251

zeaxanthin, **245**
zeolite, 152
zinc citrate, 100, 102
zinc oxide, 180, 239
zirconium chlorohydrate, 260

"This book does perfectly what it was intended to do. To provide a warm welcome and helpful support to the novices, giving them all they need to understand the scientific, commercial and regulatory aspects of cosmetics and the industry. But it does more. Anyone who has recently become involved in cosmetics will find it just as helpful learning and gaining insight into disciplines they are not particularly familiar with. The experts will find this to be an excellent teaching material. The book is comprised of cohesive chapters, yet chapters of particular interest can be read individually to make perfect sense. Overall, I cannot say enough good things about this book; a 'must-have' item for anyone of any level interested in cosmetic science and the industry. Thoroughly recommended."

Fujihiro Kanda,
Mukogawa Women's University, Japan

"I don't know a book like this on the planet; and while I have written much about cosmetic science, and edited considerably to help other high-power experts write what they have to say more clearly, all of that is for the cosmetic scientists who delve far more deeply into this fascinating topic than we-here-now. In my opinion, this book goes beyond all that as it is intended for YOU, the interested layperson who uses all these products in one form or another. Those of you who want to feel good about yourself, look good to others, have the feeling of well-being that we all seek, and be "Beautiful" no matter how old we are, what travels/travails life has brought to us, or paths we have taken without knowing where they will lead. Thus, as you go through this "must-read" short book, do as the poet Robert Frost once said: take the Road Less Travelled… and it will make all the difference."

Meyer R. Rosen
Editor in Chief: Harry's Cosmeticology, 9th edn and
Eurocosmetics Magazine

"Based upon my 40+ years in the cosmetic industry, serving for the last 4 years as the education chair of International Federation of the Society of Cosmetic Chemists, Discovering Cosmetic Science brings a much needed clarity of thought and presentation to cosmetic science. The fact that people of all levels of experience and education can read, enjoy and easily learn cosmetic science make this book an essential for everyone that makes, uses or sells cosmetics!"

Tony O'Lenick
Nascent Technologies Corporation